Modelling of Complex Signals in

Jüri Engelbrecht · Kert Tamm · Tanel Peets

Modelling of Complex Signals in Nerves

 Springer

Jüri Engelbrecht
Institute of Cybernetics
Tallinn University of Technology
Tallinn, Estonia

Kert Tamm
Department of Cybernetics
Tallinn University of Technology
Tallinn, Estonia

Tanel Peets
Institute of Cybernetics
Tallinn University of Technology
Tallinn, Estonia

ISBN 978-3-030-75041-1   ISBN 978-3-030-75039-8   (eBook)
https://doi.org/10.1007/978-3-030-75039 8

Mathematics Subject Classification: 92C05, 92C20, 35Q92, 74J35, 97M10

This Springer imprint is published by the registered company Springer Nature Switzerland AG
The registered company address is: Gewerbestrasse 11, 6330 Cham, Switzerland

Dedicated to our families

Preface

This book is about a fascinating problem in biophysics – modelling of signals in nerves. In other words, it is about casting the results of experiments into mathematical language. The starting point is physics with its conservation laws and classical equations of mathematical physics. Whatever are the governing equations for nerve signals, either the classical ones or recently derived, these are deduced from basic equations describing electrical and mechanical dynamical processes and heat conduction. The modifications of these equations used for describing physiological effects are based on experimentally observed phenomena. Such an approach means working at the interface of physics, continuum mechanics, thermodynamics and physiology. The mathematical model described in this book is a system of differential equations united into a whole by coupling forces. In building up such a system, several hypotheses are made based on physical considerations concerning the equations and coupling forces. The main aim is to describe fundamental physiological effects as well as leaving the door open for possible modifications.

The background of the authors is related to studies of wave motion in solids, biotissues and fluids characterised by keywords like solitons, dispersion, nonlinearities, microstructures, etc. The early studies in deriving the evolution equations for an electrical signal in a nerve have paved the road to the analysis of complexity of processes in nerves. Indeed, the coupling of effects in nerves resulting in an ensemble of waves is a clear sign of complexity as understood nowadays in many fields of science including biology.

The book is based on earlier publications of the authors during the last decade. Certainly, the efforts are made to build a whole. It means that physical background is described as well as some ideas from continuum mechanics used for deriving the governing equations. In several cases, the mathematical analysis is also presented in detail if the equations could be used more widely than just for the analysis of signals in nerve. For example, this is a case of longitudinal waves in a lipid bilayer (biomembrane). But without any doubt, the modelling must be consistent with experimental results like the basic conservation laws are consistent.

The studies described in this book are certainly influenced by the recent research results in this field. In this sense, the Workshops on the Physics of Excitatory

Mechanics (POEM in Copenhagen and Bad Honnef) have been extremely useful and the authors would like to thank the colleagues of this invisible college for fruitful discussions. This book could not have been written without the support of the Department of Cybernetics of Tallinn University of Technology and the former Centre for Nonlinear Studies (CENS) within this Department.

One should always indicate the funding organisations - this research was supported by the EU through the European Regional Development Fund and by the Estonian Research Council (ETAg) through various funding schemes.

We would like to thank the anonymous reviewers whose very constructive comments helped us to polish the manuscript, enlarge the lists of references and strengthen our explanations.

We appreciate very much the assistance of Martin Peters and Leonie Kunz from Springer-Verlag for producing the book.

Our most sincere thanks go to our families for their understanding and love.

Tallinn, Estonia,   *Jüri Engelbrecht*
November 2020   *Kert Tamm*
                                                                         Tanel Peets

Acknowledgements

We are grateful to the following publishers and journals for granting us their kind permission to reproduce the figures and excerpts from texts (with All Rights Reserved):

- To Elsevier (©Elsevier) from:

  - Iwasa, K., Tasaki, I.: Mechanical changes in squid giant axons associated with production of action potentials. Biochem. Biophys. Res. Commun. **95**(3), 1328–1331 (1980). DOI 10.1016/0006-291X(80)91619-8
  - Tasaki, I., Kusano, K., Byrne, P.M.: Rapid mechanical and thermal changes in the garfish olfactory nerve associated with a propagated impulse. Biophys. J. **55**(6), 1033–1040 (1989). DOI 10.1016/S0006-3495(89)82902-9
  - Peets, T., Tamm, K., Simson, P., Engelbrecht, J.: On solutions of a Boussinesq-type equation with displacement-dependent nonlinearity: A soliton doublet. Wave Motion **85**, 10–17 (2019). DOI 10.1016/j.wavemoti.2018.11.001

- To the Estonian Academy of Sciences (©Estonian Academy of Sciences) from:

  - Engelbrecht, J., Tamm, K., Peets, T.: On mechanisms of electromechanophysiological interactions between the components of nerve signals in axons. Proc. Estonian Acad. Sci. **69**(2), 81–96 (2020). DOI 10.3176/proc.2020.2.03.

- To John Wiley and Sons (©John Wiley and Sons) from:

  - Hodgkin, A.L., Katz, B.: The effect of temperature on the electrical activity of the giant axon of the squid. J. Physiol. **109**(1-2), 240–249 (1949). DOI 10.1113/jphysiol.1949.sp004388
  - Terakawa, S.: Potential-dependent variations of the intracellular pressure in the intracellularly perfused squid giant axon. J. Physiol. **369**(1), 229–248 (1985). DOI 10.1113/jphysiol.1985.sp015898

- To Springer Nature (©Springer Nature) from:

  - Hodgkin, A.L., Huxley, A.F.: Action potentials recorded from inside a nerve fibre. Nature **144**(3651), 710–711 (1939). DOI 10.1038/144710a0

– Engelbrecht, J., Peets, T., Tamm, K.: Electromechanical coupling of waves in nerve fibres. Biomech. Model. Mechanobiol. **17**(6), 1771–1783 (2018). DOI 10.1007/s10237-018-1055-2
– Engelbrecht, J., Tamm, K., Peets, T.: Internal variables used for describing the signal propagation in axons. Contin. Mech. Thermodyn. **32**, 1619–1627 (2020). DOI 10.1007/s00161-020-00868-2
– Peets, T., Tamm, K.: Mathematics of nerve signals. In: A. Berezovski, T. Soomere (eds.) Applied Wave Mathematics II, Mathematics of Planet Earth, vol. 6, pp. 207–238. Springer, Cham (2019). DOI 10.1007/978-3-030-29951-4_10

• To Taylor & Francis Group (©Taylor & Francis) www.tandfonline.com from:

– Engelbrecht, J., Tamm, K., Peets, T.: On solutions of a Boussinesq-type equation with displacement-dependent nonlinearities: the case of biomembranes. Philos. Mag. **97**(12), 967987 (2017). DOI 10.1080/14786435.2017.1283070

• To Walter de Gruyter and Company (©Walter de Gruyter and Company) from:

– Tamm, K., Engelbrecht, J., Peets, T.: Temperature changes accompanying signal propagation in axons. J. Non-Equilibrium Thermodyn. **44**(3), 277–284 (2019). DOI 10.1515/jnet-2019-0012; permission conveyed through Copyright Clearance Center, Inc.

All sources are indicated in the figure captions as well as by references in the text.

Contents

Part III Modelling of Dynamical Physiological Processes

Chapter 1
Introduction

*I suggest that the next stage in the development of biological
science will be revolutionary in its conceptual foundations and
strongly mathematical in its methods.*

Denis Noble, 2010

Mankind has always been interested in consciousness. It is certainly not only a question for philosophers but most of all for neuroscientists and is called sometimes 'the hard problem'. The mental phenomena, processing of neural information, cognitive processes, governing of functions in living bodies, etc., are all intensively studied. To understand the neural responses of the brain to various stimuli one should also understand the physical mechanisms how signals propagate in a huge network of neurons. This has created many areas of inquiries involving also a question of how patterns of neuron activities reflect the conscious state. Leaving aside the myriad of questions on consciousness, neuroscience tries to answer the question of how signals propagate along nerves. Much is known about signal propagation in nerves thanks to many experimental studies and theoretical predictions over the last two centuries but answered questions usually generate new questions. This goes along the ideas of S. Firestein [17]:

> One good question can give rise to several layers of answers, can inspire decades-long searches for solutions, can generate whole new fields of inquiry, and can prompt changes in entrenched thinking. Answers, on the other hand, often end the process.

The question we ask within this book is the following: Is it possible to construct a unified model for the propagation of a signal along a nerve where all (or at least significant) possible electrical, mechanical and thermal effects are taken into account? This seems to be a hot topic especially during the last decade with many excellent ideas and ingenious experiments (see references later).

One is clear – the propagation of a nerve impulse is a complex process spiced with nonlinearities. The present understanding of complexity started with ideas in the mid-20th century and its ideas and methods are now widely used in many branches of scientific inquiry [15, 30, 32, etc.]. In a nutshell, complex systems involve many constituents (components) and are characterised by interactions of their components. These interactions bring about a new quality at the macrolevel. Here we leave aside how complex dynamics affects consciousness [24] and focus on the modelling of complex effects in single nerve fibres. Such modelling calls for analysis at the interface of physiology, physics and mathematics. It means casting the physiological

© The Author(s), under exclusive license to Springer Nature Switzerland AG 2021
J. Engelbrecht et al., *Modelling of Complex Signals in Nerves*,
https://doi.org/10.1007/978-3-030-75039-8_1

phenomena and functional behaviour (i.e., physics) into the language of mathematics. Quite certainly, this is not an original idea. One could recall the saying of Galileo Galilei: The Book of Nature is written in the language of mathematics. Later Albert Einstein [9] had a similar opinion:

> ... mathematical construction enables us to discover the concepts and laws connecting them which give us the key to the understanding of the phenomena of Nature.

However, one cannot forget the physiology and experiments to justify the mathematical constructions.

The history of studies in neuroscience is full of remarkable observations and ideas. So Luigi Galvani (1737-1798) with his wife Lucia have conducted the experiments with frogs and discovered so-called animal electricity (the second half of the 18th century). The new idea was that the activation of muscles was generated by an electrical fluid in nerves. Nowadays we know such studies under the name 'electrophysiology'. Further on, during the 19th-century electrophysiology has been developed and the names of Emil du Bois-Reymond (1818-1896) and Julius Bernstein (1838-1917) are worth to be mentioned. Du Bois-Reymond has described the variation of signal in a nerve which in contemporary terms is called the action potential. Bernstein has described the differences in ion concentrations inside cells compared with concentration outside cells using the Nernst equation. But one cannot forget that Hermann von Helmholtz (1821-1894) has measured the velocity of the signal propagating in a nerve fibre of a frog (27.4 m/s).

Starting from the beginning of the 20th century, the history of electrophysiology is full of remarkable results related to the propagation of nerve impulses. The list of Nobel laureates in physiology or medicine is impressive. Here we just list the names (see Nobel Assembly at Karolinska Institutet):

1906: Camillo Golgi, Santiago Ramon y Cajal – in recognition of their work on the structure of the nervous system;

1932: Sir Charles Scott Sherrington, Edgar Douglas Adrian – for their discoveries regarding the functions of neurons;

1936: Sir Henry Hallett Dale, Otto Loewi – for their discoveries relating to chemical transmission of nerve impulses;

1944: Joseph Erlanger, Herbert Spencer Gasser – for their discoveries relating to the highly differentiated functions of single nerve fibres;

1963: Sir John Carew Eccles, Alan Lloyd Hodgkin, Andrew Fielding Huxley – for their discoveries concerning the ionic mechanisms involved in excitation and inhibition in the peripheral and central portions of the nerve cell membrane;

1970: Sir Bernard Katz, Ulf von Euler, Julius Axelrod – for their discoveries concerning the humoral transmitters in the nerve terminals and the mechanism for their storage, release and activation;

1991: Erwin Neher, Bert Sackmann – for their discoveries concerning the function of single ion channels in cells;

2000: Arvid Carlsson, Paul Greengard, Eric R. Kandel – signal transduction in the nervous system.

Many more names must be mentioned: B. van der Pol, K.F. Bonhoeffer, A.V. Hill, R. FitzHugh, J. Nagumo, I. Tasaki, I. Iwasa, etc. who also paved the road by their theoretical or experimental studies for contemporary understandings. The reader is referred to overviews of electrophysiology which describe these results in detail [4, 29, 33] and pay also attention to the links with mathematics [16, 18, 38].

Despite all these excellent results, there is a need for a unifying theory of nerve impulse propagation with accompanying significant effects. It is quite interesting to collect the ideas of several researchers related to this idea.

Hodgkin [20] has said:

> In thinking about the physical basis of the action potential perhaps the most important thing to do at the present moment is to consider whether there are any unexplained observations which have been neglected in an attempt to make the experiments fit into a tidy pattern.

Andersen et al. [1] have analysed several models and in this context stressed that there is a need

> to frame a theory that incorporates all observed phenomena in one coherent and predictive theory of nerve signal propagation.

Drukarch et al. [8] have recently reflected the contemporary insights in modelling of nerve impulses and concluded that this field could lead to

> the formulation of a more extensive and complete conception of the nerve impulse.

Consequently, the community of researchers have seen the shortcomings of theories and explanations and called for further studies in order to explain the measured physical phenomena more coherently.

In what follows in this book, is an overview of mathematical modelling of the fascinating problem of nerve pulse propagation. It is clear that such modelling stands "on the shoulders of Giants" as Isaac Newton said already in 1675. The list of Nobel laureates (see above) is the best proof for this statement.

Every mathematical modelling starts with the description of reality – this was known already to Leonardo da Vinci: observe the phenomenon and list quantities having a numerical magnitude that seems to influence it (cited after Truesdell [37]). The nerve impulse propagation is a complex dynamical process in the nerve fibre structured as a cylindrical tube embedded in a certain environment. The tube is filled with a fluid called axoplasm and has a wall called biomembrane. According to the Hodgkin-Huxley paradigm (ionic hypothesis), an electrical signal called action potential (AP) is generated in the axoplasm due to complicated mechanisms of changes in ion concentrations (ion currents). The propagation of the AP is accompanied by mechanical and thermal effects in the axoplasm and biomembrane with possible feedback from them to the AP. This description is certainly only the backbone of the complicated process but sufficient to envisage the ideology of mathematical modelling.

The modelling starts by describing the physics of physiological processes followed by the mathematical descriptions of phenomena. The complex process is built up by single constituents – the physics of single structural elements. First one should understand the behaviour of those single elements and then unite them into a whole. In doing so, we follow the warning of Toffler [36]:

> One of the most highly developed skills in contemporary Western civilisation is dissection: the split-up of problems into their smallest possible components. We are good at it. So good, we often forget to put the pieces back together again.

We aim to put the pieces back together by coupling and certainly one cannot forget the verification of results which is a must for all mathematical modelling [11]. In doing so, one should start from the main principles of complexity, with special attention to the biological world. The analysis of the basic models (equations) of mathematical physics follows to demonstrate the basic rules and needs for modifications for grasping the reality. One should distinguish between the macro- and microscales. If we consider the structural elements of nerve fibres at the macroscale then their microstructure needs to be taken into account by certain modifications of governing equations. Although here we deal with the biological structures, the knowledge from continuum mechanics is of the great help in modelling the coupling effects.

As a result, a system of differential equations is constructed which governs the electrical, mechanical and thermal effects in nerve fibres. The single equations are all related to the known mathematical models but the coupling terms including the possible feedback are proposed based on the leading hypothesis: the accompanying mechanical and thermal effects are generated due to changes in electrical signals (AP and/or ion currents) and these couplings are described by coupling forces [13]. The system of governing equations is then solved numerically and results compared with experimental data.

Such a phenomenological approach described above follows the ideas of integrative modelling in biology: from descriptive to integrative level and then explanatory level to gain physiological insight [31]. In other words, it is integrative biological modelling *in silico* [26]. In more specific terms, it is modelling of processes at the interface of physiology, physics and mathematics [2, 14] with the special attention to electromechanophysiological interactions in nerve fibres. In more general terms, it is the modelling of biological complexity [6].

The first part of the book explains the importance of complexity in biological processes and the possibilities of mathematical modelling in terms of mathematical physics. Next, the physiology of axons is described and the experimental results analysed which constitute the basis for further mathematical modelling. It means that physical mechanisms are described mathematically for building a general mathematical model for signal propagation in nerves. The computational simulations have demonstrated the properties of solutions of single model equations as well as of the coupled system resulting in an ensemble of waves.

Chap. 2 – Complexity. Here the background of the analysis is presented. Without any doubt, the signal propagation in nerves is a complex process with many interacting components. That is why the main features of complexity science are briefly presented. The complexity of physical processes is described [15, 30] and the typical effects and results are explained. Within the context of this treatise, the attention is also paid to biological complexity [6, 39] characterised by the coupling of many processes of the different origin and multiple scales. This leads to the need for the mathematical modelling of complex processes which in the framework of biology is called the computational (*in silico*) biology [31]. The importance of such an approach is analysed in details by the National Research Council [28]. This report says:

> ... a mathematical model can highlight basic conceptions and identify key factors or components of a biological system. In addition, models enable to formalize the intuitive understandings and "link what is known to what is yet unknown".

Such a description fully corresponds to the ideology of this book.

Chap. 3 – Waves. The signal propagation means physically wave motion. Whatever the medium is, the background for such processes is physical and related to the basic wave equation. This is one of the equations that "changed the world" [35]. As one of the basic equations of mathematical physics, usually it must be modified to model the realities better. Another basic equation is the diffusion equation which like wave equation gives rise to many modified models. There are many physical effects that must be accounted for a description of processes: forcing, nonlinear and dispersive effects, the influence of inhomogeneities, dissipative effects, etc. By using modified presentations, it is possible to describe many types of waves including solitons and solitary waves.

These Chapters on complexity and waves describe the background of the problem in physical and mathematical terms and the next Chapters are devoted to the modelling of dynamical physiological processes.

Chap. 4 – Nervous signals. The propagation of the action potential (AP) in axons is well characterised in many experimental and theoretical studies [3, 5, 7]. After the benchmark studies [20], many details responsible for excitability of the AP, ionic mechanisms (voltage-gated ion channels in the biomembrane), biophysical forces, the effects of anaesthesia, etc. are known nowadays. The classical models are not always sufficient to grasp such details which are signs of the high complexity and coupling of many processes. The general description of axon physiology in this Chapter prepares the ground for the analysis of dynamics in the structural elements of axons.

Chap. 5 – Dynamical effects in nerves. The experimental results permit to get the general picture of signals in nerve fibres. This concerns first of all the propagation of the AP which is accompanied by other dynamical effects much in the sense of complexity analysed above in Chap. 2. The biomembrane, made of lipids is a deformable medium and consequently may deform during the passage of the signal in a nerve [19]. By analogy to continuum mechanics, more specifically to the theory of rods, the longitudinal and transverse deformations of the cylindrical biomembrane are coupled. The AP generates also experimentally measured pressure wave in the axoplasm and the temperature change. The general overview of all these effects is the basis for mathematical models constituting the **wave ensemble**. The assumptions for constructing such an ensemble are presented. The **main hypothesis** for the source of coupling forces is formulated: the mechanical waves in the biomembrane and the axoplasm, as well as temperature changes, are generated due to changes in electrical signals (the AP and/or ion currents) with possible feedback [13].

Chap. 6 – Mathematics of single effects. Before deriving a coupled mathematical model, the models for the single elements must be understood. The mathematical models for the AP are presented, starting from the celebrated Hodgkin-Huxley (HH) model [21] followed by its simplified variant called FitzHugh-Nagumo (FHN) model [27]. However, there are also some other simplified models which have been derived to pay attention to specific details [10]. The longitudinal waves in biomembranes (lipid bi-layers) are governed by the Boussinesq-type equation involving nonlinear and dispersive effects [12, 19]. The pressure wave in the axoplasm may be described by either the Navier-Stokes equation or by a modified wave equation. All these processes are governed by wave-like equation. The temperature is governed by a diffusion-type equation while the transverse displacement of the biomembrane is calculated from the longitudinal deformation. The analysis of these single mathematical models opens the richness of the processes.

Chap. 7 – Physical mechanisms. Here the electromechanophysiological interactions between the components of nerve signals are described based on experimental studies. Three basic physical mechanisms responsible for interactions are: (i) electric-lipid bilayer interaction resulting in the mechanical wave in the biomembrane; (ii) electric-fluid interaction resulting in the mechanical wave in the axoplasm; (iii) electric-fluid interaction resulting in the temperature change in the axoplasm. Experimental results are analysed and the possible mathematical description is presented for building up the mathematical model for coupling. The concept of internal variables [25] is used for describing the influence of microdynamics responsible for temperature effects.

After the analysis of physical effects in nerve fibres and their possible mathematical description, the further step is to collect all this knowledge into the building up a coupled model for signal propagation in axons. This means putting the pieces back to the whole. In this context, one should also stress the ideas of Systems Biology. Kohl et al. [23] said:

> Systems Biology is an approach to biomedical research that consciously combines reduction and integration of information across multiple spatial scales to identify and characterise parts and explore the ways in which their interaction with one another and with the environment results in the maintenance of the entire system.

Chap. 8 – An ensemble of waves. It is worth to remember the statement of Kaufmann [22]:

> Electrical action potentials are inseparable from force, displacement, temperature, entropy and other membrane variables.

The whole signal in a nerve is an ensemble which includes primary and secondary components. The **primary components** are governed by wave-type equations characterised by corresponding velocities. These components are the AP, the longitudinal wave in the biomembrane and the pressure wave in the axoplasm. The **secondary components** have no specific velocities: the temperature changes and the transverse displacement of a biomembrane. The temperature change is governed by a diffusion-type equation while the transverse displacement is calculated from the longitudinal deformation of the biomembrane. The governing equations are coupled by coupling forces. The **main hypothesis** for constructing the coupling forces is formulated in Chap. 5. The energetical balance of the whole process is briefly analysed. Attention is paid to possible simplifications and modifications of the model. As far as the modelling has been presented in the dimensionless form, the possible models in physical units are also presented with the analysis of units of coupling forces.

Chap. 9 – *In silico* experiments. The full coupled model and its simplified variants for describing an **ensemble of waves** are analysed by numerical simulation. This concerns solving a system of partial differential equations which describes more or less the important physical effects related to the propagation of the AP, longitudinal wave LW in the biomembrane, pressure wave in the axoplasm PW, transverse displacement of the biomembrane TW, and temperature change Θ, which all are coupled. The proposed coupling forces involve also possible feedback. The nondimensional form of this system is solved numerically by using the pseudospectral method. The ensemble is generated by an input of an electrical spark which generated the AP and by coupling all the other components of the ensemble. The *in silico* experiments cover a wide range of examples and demonstrate good qualitative correspondence to experimental results.

Chap. 10 – Final remarks. The step-by-step analysis from basic physics over a description of effects for building a coupled mathematical model of the propagation of signals in axons is summarised. This robust model is an attempt to build up a unified model for describing all the essential effects which compose an ensemble of waves. The novelties of the model are briefly described. Within this framework, the mathematical model for longitudinal waves in the biomembrane is improved and its possible solutions analysed in detail. The basic hypothesis is that the structures of coupling forces between the components of the ensemble depend on changes in the field variables. The models for the pressure wave in the axoplasm and temperature change are proposed. The concept of internal variables is used for modelling the influence of endothermic processes in nerve fibres. The pro's and contra's of such an analysis are discussed and possible open problems for further development in modelling envisaged. The results permit to formulate general principles for modelling complex biological processes.

Two technical appendices are included. Appendix A describes the pseudospectral method used for numerical calculations, Appendix B is the source for scripts for numerical integration (in Python) and for visualisation of the wave ensemble (in Matlab) with an example.

The highlights of the presentation are the general philosophy of modelling of complex biological processes, the casting of physical effects into the mathematical language, the detailed analysis of governing equations of single effects, and the analysis of a coupled system. As a result, the robust mathematical model for signal propagation in nerves is a part of the complex nonlinear world in terms of Scott [34].

The material presented in this book can be of interest for graduate students and researchers interested in physiology and mathematical modelling.

References

1. Andersen, S.S.L., Jackson, A.D., Heimburg, T.: Towards a thermodynamic theory of nerve pulse propagation. Prog. Neurobiol. **88**(2), 104–13 (2009). DOI 10.1016/j.pneurobio.2009.03.002

2. Bialek, W.: Perspectives on theory at the interface of physics and biology. Reports Prog. Phys. **81**(1), 1–21 (2018). DOI 10.1088/1361-6633/aa995b

3. Clay, J.R.: Axonal excitability revisited. Prog. Biophys. Mol. Biol. **88**(1), 59–90 (2005). DOI 10.1016/j.pbiomolbio.2003.12.004

4. Cole, K.S.: Membranes, Ions and Impulses. A Chapter of Classical Biophysics. University of California Press, Berkley, CA (1968)

5. Courtemanche, M., Ramirez, R.J., Nattel, S.: Ionic mechanisms underlying human atrial action potential properties : insights from a mathematical model. Am. J. Physiol. **275**(1), H301–H321 (1998)

6. Coveney, P.V., Fowler, P.W.: Modelling biological complexity: a physical scientist's perspective. J. R. Soc. Interface **2**(4), 267–280 (2005). DOI 10.1098/rsif.2005.0045

7. Debanne, D., Campanac, E., Bialowas, A., Carlier, E., Alcaraz, G.: Axon physiology. Physiol. Rev. **91**(2), 555–602 (2011). DOI 10.1152/physrev.00048.2009.

8. Drukarch, B., Holland, H.A., Velichkov, M., Geurts, J.J., Voorn, P., Glas, G., de Regt, H.W.: Thinking about the nerve impulse: A critical analysis of the electricity-centered conception of nerve excitability. Prog. Neurobiol. **169**, 172–185 (2018). DOI 10.1016/j.pneurobio.2018.06.009

9. Einstein, A.: On the method of theoretical physics. Philos. Sci. **1**(2), 163–169 (1934)

10. Engelbrecht, J.: On theory of pulse transmission in a nerve fibre. Proc. R. Soc. A Math. Phys. Eng. Sci. **375**(1761), 195–209 (1981). DOI 10.1098/rspa.1981.0047

11. Engelbrecht, J.: Questions About Elastic Waves. Springer International Publishing, Cham (2015). DOI 10.1007/978-3-319-14791-8

12. Engelbrecht, J., Tamm, K., Peets, T.: On mathematical modelling of solitary pulses in cylindrical biomembranes. Biomech. Model. Mechanobiol. **14**, 159–167 (2015). DOI 10.1007/s10237-014-0596-2

13. Engelbrecht, J., Tamm, K., Peets, T.: Modeling of complex signals in nerve fibers. Med. Hypotheses **120**, 90–95 (2018). DOI 10.1016/j.mehy.2018.08.021

14. Engelbrecht, J., Tamm, K., Peets, T.: Modelling of processes in nerve fibres at the interface of physiology and mathematics Biomech. Model. Mechanobiol. (2020). Online from 4 June 2020. DOI 10.1007/s10237-020-01350-3

15. Érdi, P.: Complexity Explained. Springer, Berlin and Heidelberg (2008)

16. Ermentrout, G.B., Terman, D.H.: Mathematical Foundations of Neuroscience. Springer, New York, NY (2010). DOI 10.1007/978-0-387-87708-2

17. Firestein, S.: Ignorance. How it Drives Science. Oxford University Press, Oxford et al. (2012)

18. Franzone, C.P., Pavarino, L.F., Scacchi, S.: Mathematical Cardiac Electrophysiology. Springer International Publishing, Cham (2014). DOI 10.1007/978-3-319-04801-7

19. Heimburg, T., Jackson, A.D.: On soliton propagation in biomembranes and nerves. Proc. Natl. Acad. Sci. USA **102**(28), 9790–5 (2005). DOI 10.1073/pnas.0503823102

20. Hodgkin, A.L.: The Conduction of the Nervous Impulse. Liverpool University Press (1964)

21. Hodgkin, A.L., Huxley, A.F.: Resting and action potentials in single nerve fibres. J. Physiol. **104**, 176–195 (1945)

22. Kaufmann, K.: Action Potentials and Electromechanical Coupling in the Macroscopic Chiral Phospholipid Bilayer. Caruaru, Brazil (1989)

23. Kohl, P., Crampin, E.J., Quinn, T.A., Noble, D.: Systems biology: An approach. Clin. Pharmacol. Ther. **88**(1), 25–33 (2010). DOI 10.1038/clpt.2010.92

24. Mainzer, K.: Thinking in Complexity. The Complex Dynamics of Matter, Mind, and Mankind. Springer, Berlin et al. (1997)

25. Maugin, G.A.: Internal variables and dissipative structures. J. Non-Equilibrium Thermodyn. **15**(2) (1990). DOI 10.1515/jnet.1990.15.2.173

26. McCulloch, A.D., Huber, G.: Integrative biological modelling *In Silico*. In: G. Bock, J.A. Goode (eds.) *'In Silico'* Simulation of Biological Processes, pp. 4–25. John Wiley & Sons, Chichester (2002). DOI 10.1002/0470857897.ch2.
27. Nagumo, J., Arimoto, S., Yoshizawa, S.: An active pulse transmission line simulating nerve axon. Proc. IRE **50**(10), 2061–2070 (1962). DOI 10.1109/JRPROC.1962.288235
28. National Research Council: Catalyzing Inquiry at the Interface of Computing and Biology. The National Academies Press, Washington (2005). DOI 10.17226/11480
29. Nelson, P.C., Radosavljevic, M., Bromberg, S.: Biological Physics: Energy, Information, Life. W.H. Freeman and Company, New York, NY (2003)
30. Nicolis, G., Nicolis, C.: Foundations of Complex Systems. World Scientific, New Jersey et al. (2007)
31. Noble, D.: Chair's Introduction. In: G. Bock, J.A. Goode (eds.) *'In Silico'* Simul. Biol. Process. Novartis Found. Symp. 247, Vol. 247, pp. 1–3. Novartis Foundation (2002). DOI 10.1002/0470857897.ch1
32. Prigogine, I., Stengers, I.: Order Out of Chaos. Heinemann, London (1984)
33. Scott, A.C.: Neuroscience: A Mathematical Primer. Springer Science & Business Media, New York (2002)
34. Scott, A.C.: The Nonlinear Universe. Chaos, Emergence, Life. The Frontiers Collection. Springer, Berlin, Heidelberg (2010)
35. Stewart, I.: In Pursuit of the Unknown: 17 Equations That Changed the World. Profile Books, London (2013)
36. Toffler, A.: Foreword to: Prigogine I, Stengers I. Order out of Chaos. Heinemann, London. (1984)
37. Truesdell, C.: Essays in the History of Science. Springer, New York et al. (1968)
38. Tveito, A., Lines, G.T.: Computing Characterizations of Drugs for Ion Channels and Receptors Using Markov Models. Springer International Publishing, Cham (2016). DOI 10.1007/978-3-319-30030-6
39. Weiss, J.N., Qu, Z., Garfinkel, A.: Understanding biological complexity: lessons from the past. FASEB J. **17**(1), 1–6 (2003). DOI 10.1096/fj.02-0408rev

Part I
Complexity and Waves

Chapter 2
Complexity

*Complexity and nonlinearity are prominent features in the
evolution of matter, life, and human society.*

Klaus Mainzer, 1994

The World around us is complex and during the last half a century, much attention
is paid to the analysis of complex systems. The main reason for this interest is
that the complex systems are characterised by interactions between their elements
(constituents) and as a result, the system behaves differently from the simple sum of
the behaviours of its elements. In order to build up the background for modelling of
processes in nerves, a brief overview is presented on the essence of complexity. It is
based on earlier overviews of one of the authors [6, 7, 8, 28].

The present understanding of complexity started from ideas of L. van Bertalanffy
and N. Wiener (mid 20th-century) and developed in studies on self-organisation,
chaos theory, networks, etc. in the second half of the 20th century [1, 9, 21, 25, etc.].
In this Chapter basic ideas on physical and biological complexity are described. These
ideas have a great impact on general thinking, however, the extremely interesting
area on the philosophy of complexity is not touched here. There are many studies
by Edgar Morin, Paul Cilliers, et al. who have analysed complex thinking (see, for
example, [6]).

2.1 Complexity of Physical Systems

Contemporary ideas how to 'tame' complexity stem from studies in physics, par-
ticularly in mechanics. This is partly related to the historical development of the
scientific inquiry but partly also to an essential property of mechanical systems –
nonlinearity. Already Isaac Newton introduced the inverse-square law of gravitation
and later Henri Poincaré showed the complications which arose in the analysis of the
three-body system. Nowadays it seems clearly understood that nonlinear interactions
may cause effects that are not simply described by summation.

Let us list the essential properties of complex systems [6]:

(i) non-additivity and nonlinear interactions. This is the source for chaotic motions and typical for many physical systems modelled by mappings or differential
equations. A typical example of a nonlinear interaction is the gravitational force

between different masses. The three-body system (Sun, Earth, Moon) analysed by H. Poincaré already more than a century ago has revealed the ideas of possible instabilities. Another iconic example is the Lorenz attractor describing simplified atmospheric motion using the system of three nonlinear differential equations.

(ii) deterministic unpredictability. The behaviour of deterministic nonlinear systems may not be predicted and lead to the chaotic regimes of motion. A typical example is a simple logistic equation (mapping) derived for calculation of changes in the number of animals in a certain species. The weather is described by nonlinear Navier-Stokes equations that again do not permit the accurate forecasts for longer periods.

(iii) sensitivity to initial conditions. Small changes in initial conditions for a dynamical nonlinear process may lead to large changes in the resulting quantities with time. This phenomenon within the framework of a nonlinear simple model was discovered by Lorenz although Maxwell has already hinted to such a possibility in the 19th century and Poincaré at the beginning of the 20th century.

(iv) there are several typical phenomena characterising the behaviour of nonlinear systems like *bifurcations* when the new solutions emerge after small changes of control parameters, *emergence* when new patterns arise at the system level not predicted by fundamental proprieties of the system's constituents; *attractors* where the solutions are attracted to a certain space of variables (phase space), *multiple equilibria* which are characterised by several (co-existing) attractors, *thresholds* which mark the borders between the various states, *coherent states* where effects are balanced, *adaptability* when independent constituents interact changing their behaviours in reaction to those of others, and adapting to a changing environment; *self-organising criticality* when a complex system may possess a self-organising attractor state that has an inherent potential for abrupt transitions of a wide range of intensities while the magnitude of the next transition is unpredictable, phase transitions, etc.

(v) despite the variety of chaotic motions, there are several rules which govern the processes: period-doubling and Feigenbaum numbers, power laws, self-similarity, fractality of attractors, etc. and also several which allow to analyse the processes: Melnikov method, renormalisation method, determination of the Kolmogorov entropy and Lyapunov exponents for determining the scale of chaotic motions, etc.

This is certainly just a brief description of physical complexity: properties, effects, rules. One should also mention the structures of complex systems where studies on networks and hierarchies have demonstrated many properties indicated above. For more detailed information one should consult the Encyclopedia of Nonlinear Science [26].

Leaving aside the technical details two issues should be underlined. First, contrary to the usual understandings (common sense) the world around us is deeply nonlinear and linear models, as a rule, are simplifications. Second, due to interactions, new quality could emerge in complex systems that cannot be realised by simple summation of the single processes. This property was known a long time ago:

The whole is more than the sum of its parts.

Aristotle, Metaphysica

2.2 Complexity in Biology

Contemporary biology pays a lot of attention to understanding the biological processes on the system level. It means the analysis of system structures and system dynamics [15] and in this sense, the complexity problems are in focus like for physical systems (Sect. 2.1). Noble [22] stated,

> that in order to unravel the complexity of biological processes we need to model in an integrative way at all levels: gene, protein, pathways, subcellular, cellular, tissue, organ, system.

Without any doubt, this means an interdisciplinary approach involving physics, chemistry, and mathematics for understanding phenomena in hierarchical biological structures which stresses the ideas of multiscale modelling. Sometimes such an approach is called bio-mathematical modelling [13], sometimes a physical scientists perspective [4], sometimes simply working at the interface of physics and biology [2].

Compared with physics, there are several issues to be taken into account [27]:
- biological systems may involve many chemical reactions and transfer mechanisms where the constituents of a system should be analysed on the molecular level;
- biological systems need energy exchange with the surrounding environment and represent often the systems far from the thermodynamic equilibrium;
- the processes operate over different time scales, are spatially extended, and include many hierarchies;
- in physical terms, one should account for dissipation, activity/excitability, spatiotemporal coupling and one cannot forget about the possible nonlinear effects (additivity is lost);
- in mathematical terms, the biological systems can often be described by different types of mathematical equations which causes difficulties in solving them.

As said already above, the hierarchies in biological systems play an important role in understanding the behaviour of a system as a whole. In general, hierarchies can be either structural or functional [19]. Structural hierarchies reflect the enormously rich architecture of biological systems and/or tissues. One could distinguish such structural hierarchies:
atom – molecule – cell – tissue – organ – human [18];
genes – mRNA – proteins – cells – tissue – organ – body [11];
sarcomeres – myofibrils – fibres – myocardium – heart [19];
etc.
These hierarchies are characterised by multiple scales. For example, from around 10^{-10} m for an atom to 1-2 m for a body. This is illustrated in Fig. 2.1.

Functional hierarchies reflect the complexity of functioning biosystems. In principle, it is possible to distinguish two types of functioning: a sequence in series (bottom-up or top-down) and a sequence with parallel connections [27]. An example of the first type is the functioning of the heart: oxygen consumption – energy transfer – Ca^{2+} signals – cross-bridge motion – contraction. An example of the second type is ischemia: electrical activity – creatine kinase system/sodium-potassium

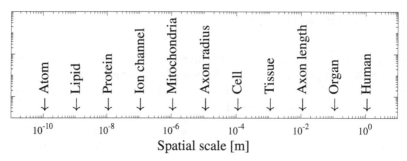

Fig. 2.1 Some scales in biological structural hierarchies. Note that logarithmic scale is used.

pump/intracellular calcium concentration – arrhythmic contracture [17]. One should note the excellent results in functionally integrated modelling of the heart [19, 23]. The modelling involves organ system model, whole heart continuum model, tissue model, single-cell model, macromolecular complex model, molecular model, atomic model – all integrated into a whole. The steps of such an integrated model are in detail analysed by McCulloch and Huber [19] over many scales. As described by Noble [24], the integrated view on interplay between genes, cells, organs, body, and the environment is metaphorically like a kind of music. In this context, there is an [13]

> increasing demand in quantitative assessment of element inter-relations in complex biological systems.

Nervous systems governing the behaviour of living species are extremely complex. Sometimes it is said that the brain is the most complex system in the universe [16] having about one hundred billion neurons and about one hundred trillion connections. Leaving aside the neuronal networks, the propagation of signals in nerve fibres from one cell to another constitute the basic elements of a neuronal system. This process is characterised by many interactions and can be taken as a basic complexity element. This book is about the analysis at that level.

2.3 Mathematical Modelling

There are several ways of understanding the phenomena in the physical world – measurements, theories and modelling. The latter includes mostly casting a real-world system or a phenomenon or a process into a mathematical representation. Physics and mathematics are closely related areas over centuries but nowadays mathematical modelling in biology is gaining more and more attention and the role of mathematical modelling is obvious. It helps to put the pieces (i.e., single phenomena) together in sense of Toffler as stressed by Noble [23].

A mathematical model is an abstraction of a real process/phenomenon presented in the language of mathematics. It means that the variables are quantified and they are described by equations or functions. In general, a model includes the governing equations, initial and boundary conditions and usually also certain constraints, all based on a certain set of assumptions. An input generates a certain output and implicitly it means causality. It is clear that mathematical modelling must be based on definite rules but on the other hand be flexible enough to grasp possible modifications if needed. In continuum mechanics, for example, so-called axioms of constitutive theory must be satisfied [10]. These axioms are: (i) causality; (ii) determinism; (iii) equipresence; (iv) objectivity; (v) time reversal; (vi) material invariance; (vii) admissibility; (viii) neighbourhood; (ix) memory. In the analysis of dynamical processes, such axioms are usually followed although some of them (neighbourhood, memory) may be modified [5].

Some mathematical models form the basic knowledge and even not thought always in mathematical terms. Take, for example, the theorem of Pythagoras:

$$a^2 + b^2 = c^2, \tag{2.1}$$

which is a part of primary education. This theorem describes the relationship between the sides (the legs and hypothenuse) in a right triangle. Newton's law of universal gravitation is another iconic model:

$$F = G\frac{m_1 m_2}{r^2}, \tag{2.2}$$

which says that a point mass m_1 attracts another point mass m_2 by a force F proportional to the product of these point masses and inversely proportional to the square of distance r between them while G is the gravitational constant.

The celebrated Albert Einstein's formula

$$E = mc^2, \tag{2.3}$$

which relates energy E, mass m and the speed of light c has a fundamental significance. The list of such iconic mathematical models is long, including, for example, the Lotka-Volterra model of predator-prey interaction, the Lorenz model of the atmospheric motion, etc., etc. These examples demonstrate vividly that through mathematical descriptions new knowledge is obtained.

However, the list of unsolved problems is endless. It is said [3]:

> To explain a phenomenon is to find a model that fits it into the basic framework of the theory and that thus allows us to derive analogues for the messy and complicated phenomenological laws which are true of it. The models serve a variety of purposes, and individual models are to be judged according to how well they serve the purpose at hand.

The general aims of modelling could be formulated in the following way:
- studying the properties of various processes or phenomena;
- obtain new knowledge.

A possible flowchart of modelling is shown in Fig. 2.2 which reflects the validation and possible iterative improvement of mathematical models.

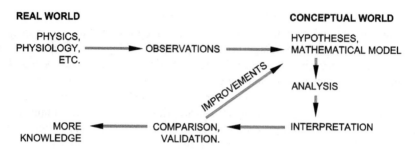

Fig. 2.2 A possible flowchart of mathematical modelling.

The various fields of scientific inquiry need various approaches. Concerning the biological phenomena, there are several issues one should take into account [27]:
- in addition to the physical, chemical and geometrical properties of biosystems being modelled, certain hierarchies of processes, substructures and variables ordered by scales must be taken into account;
- whatever the mathematical apparatus is, a model cannot reflect the behaviour of the system over all the time scales and conditions but describes usually only its basic (backbone) or specific (under special conditions) features;
- in Systems Biology, other fields of knowledge such as chemistry, thermodynamics, computational methods, etc. should be interwoven into a whole on the basis of biological functions.

So the mathematical modelling in biology should be based at the interface of physics, mathematics and biology (and probably chemistry should also be added to this list). Examples described by Gavaghan et al. [13], Bialek [2] and others demonstrate vividly how fruitful such an approach is. Activity of brain is related to mathematics, solid and fluid mechanics, biomechanics, neuropathology, neuro-surgery, etc. and the studies of brain must be based on multidisciplinary analysis [14]. Later in this book, we demonstrate how the knowledge from mathematical physics, continuum mechanics and mathematics can be used in the analysis of nerve signals. This concerns the structure of governing equations, the properties of coupling forces, internal variables to account for processes at a smaller scale, etc. Following Engel-brecht [5] we ask how to modify the governing equations in order to model physical effects which influence the process.

A Report from the National Research Council [20] includes a special Chapter "Computational Modeling and Simulation as Enablers for Biological Discovery" where the features of modelling were discussed. The list of properties and possibili-ties of models is worth to be repeated here:

models provide a coherent framework for interpreting data;
models highlight basic concepts of wide applicability;
models uncover new phenomena or concepts to explore;
models identify key factors or components of a system;
models can link levels of detail;
models enable the formalization of intuitive understandings;

models can be used as a tool for helping to screen unpromising hypotheses;
models inform experimental design;
models can predict variables inaccessible to measurement;
models can link what is known to what is yet unknown;
models can be used to generate accurate quantitative predictions;
models expand the range of questions that can be meaningfully be asked.

Later in this book, we shall try to emphasise some of these statements. Certainly, the problems of bio-mathematical modelling have been widely analysed in the community [4, 13, 19, 22]. One could agree with the National Research Council [20]:

Models are useful for formalizing intuitive understandings, even if those understandings are partial and incomplete. What appears to be a solid verbal argument about cause and effect can be clarified and put to a rigorous test as soon as an attempt is made to formulate the verbal arguments into a mathematical model. This process forces a clarity of expression and consistency (of units, dimensions, force balance, or other guiding principles) that is not available in natural language. As importantly, it can generate predictions against which intuition can be tested.

In most studies on biological complexity and mathematical modelling, the importance of interactions is stressed. This is closely related to the causality of processes. In this way, mathematical models could give valuable information about possible phenomena and *in silico* simulations enlarge essentially our understandings on biological processes.

Generally speaking, modelling has generated many debates in the philosophy of science where the crucial problem is whether the assumption and hypotheses made in modelling reflect the reality [3, 12, etc].

Finally, let us recall the quotation of Toffler in Chap. 1 about collecting the pieces into a whole. A similar idea is mentioned by Noble [22] following Lewis Carroll:

We have, successfully broken Humpty-Dumpty down into his smallest bits. Do we now have to worry about how to put him back together again?

Our answer to this question is "Yes".

References

1. Bak, P.: How Nature Works. Oxford University Press (1997)
2. Bialek, W.: Perspectives on theory at the interface of physics and biology. Reports Prog. Phys. **81**(1), 012601 (2018). DOI 10.1088/1361-6633/aa995b
3. Cartwright, N.: How the Laws of Physics Lie. Oxford University Press (1983). DOI 10.1093/0198247044.001.0001
4. Coveney, P.V., Fowler, P.W.: Modelling biological complexity: A physical scientist's perspective. J. R. Soc. Interface **2**(4), 267–280 (2005). DOI 10.1098/rsif.2005.0045
5. Engelbrecht, J.: Questions About Elastic Waves. Springer International Publishing, Cham (2015). DOI 10.1007/978-3-319-14791-8
6. Engelbrecht, J.: The knowledge of complexity should be a part of contemporary education. Eruditio **2**(4), 27–37 (2018)

7. Engelbrecht, J., Berezovski, A., Soomere, T.: Highlights in the research into complexity of nonlinear waves. Proc. Estonian Acad. Sci. **59**(2), 61–65 (2010). DOI 10.3176/proc.2010.2.01

8. Engelbrecht, J., Peets, T., Tamm, K., Laasmaa, M., Vendelin, M.: On the complexity of signal propagation in nerve fibres. Proc. Estonian Acad. Sci. **67**(1), 28–38 (2018). DOI 10.3176/proc.2017.4.28

9. Érdi, P.: Complexity Explained. Springer, Berlin and Heidelberg (2008)

10. Eringen, A.C., Maugin, G.A.: Electrodynamics of Continua I. Springer New York, New York, NY (1990). DOI 10.1007/978-1-4612-3226-1

11. Fernandez, J.W., Schmid, H., Hunter, P.J.: A framework for soft tissue and musculo-skeletal modelling: clinical uses and future challenges. In: G.A. Holzapfel, R.W. Ogden (eds.) Mechanics of Biological Tissue. Proc. IUTAM Symp., pp. 339–354. Springer, Berlin (2006). DOI 10.1007/3-540-31184-X_24

12. Frigg, R., Hartmann, S.: Models in Science. In: E. Zalta (ed.) Stanford Encycl. Philos. (2018)

13. Gavaghan, D., Garny, A., Maini, P.K., Kohl, P.: Mathematical models in physiology. Philos. Trans. R. Soc. A Math. Phys. Eng. Sci. **364**(1842), 1099–1106 (2006). DOI 10.1098/rsta.2006.1757

14. Goriely, A., Geers, M.G., Holzapfel, G.A., Jayamohan, J., Jérusalem, A., Sivaloganathan, S., Squier, W., van Dommelen, J.A., Waters, S., Kuhl, E.: Mechanics of the brain: perspectives, challenges, and opportunities. Biomech. Model. Mechanobiol. **14**(5), 931–965 (2015). DOI 10.1007/s10237-015-0662-4.

15. Kitano, H.: Systems biology : a brief overview. Science **295**(5560), 1662–1664 (2002). DOI 10.1126/science.1069492

16. Koch, C., Laurent, G.: Complexity and the nervous system. Science **284**(5411), 96–98 (1999). DOI 10.1126/science.284.5411.96

17. Kohl, P., Crampin, E.J., Quinn, T.A., Noble, D.: Systems biology: An approach. Clin. Pharmacol. Ther. **88**(1), 25–33 (2010). DOI 10.1038/clpt.2010.92

18. Kolston, P.J.: Finite-element modelling: a new tool for the biologist. Phil. Trans. R. Soc. Lond. A, **358**, 611–631 (2000). DOI 10.1098/rsta.2000.0548

19. McCulloch, A.D., Huber, G.: Integrative biological modelling *In Silico*. In: G. Bock, J.A. Goode (eds.) '*In Silico*' Simul. Biol. Process., pp. 4–25. John Wiley & Sons, Chichester (2002). DOI 10.1002/0470857897.ch2

20. National Research Council: Catalyzing Inquiry at the Interface of Computing and Biology. The National Academies Press, Washington (2005). DOI 10.17226/11480

21. Nicolis, G., Nicolis, C.: Foundations of Complex Systems. World Scientific, New Jersey et al. (2007)

22. Noble, D.: Chair's Introduction. In: G. Bock, J.A. Goode (eds.) '*In Silico*' Simul. Biol. Process. Novartis Found. Symp. 247, Vol. 247, pp. 1–3. Novartis Foundation (2002). DOI 10.1002/0470857897.ch1

23. Noble, D.: Modeling the heart – from genes to cells to the whole organ. Science **295**(5560), 1678–1682 (2002). DOI 10.1126/science.1069881

24. Noble, D.: The Music of Life: Biology Beyond Genes. Oxford University Press (2006)

25. Prigogine, I., Stengers, I.: Order Out of Chaos. Heinemann, London (1984)

26. Scott, A. (ed.): Encyclopedia of Nonlinear Science. Taylor &Francis, New York (2005)

27. Vendelin, M., Saks, V., Engelbrecht, J.: Principles of mathematical modeling and *in silico* studies of integrated cellular energetics. In: V. Saks (ed.) Molecular System Bioenergetics: Energy for Life, pp. 407–433. Wiley, Weinheim (2007). DOI 10.1002/9783527621095.ch12

28. Weiler, R., Engelbrecht, J.: The new sciences of networks & complexity: a short introduction. Cadmus J. **2**(1), 131–141 (2013)

Chapter 3
Waves

Wave is a state moving into another state.

Clifford Truesdell, Walter Noll, 1965

3.1 Preliminaries

Neuroscience is sometimes characterised as information processing in neural net-works. The basic elements of these networks are nerve fibres where the signals propagate. The analysis of these basic elements means that one should deal with dynamical processes and that brings us directly to the concept of wave motion. That is why one should start with the physics and the corresponding description of waves. The main question is how the physics of wave motion can be described in mathematical terms. As described before (see Chap. 1), a nerve fibre is composed of many components with various physical properties and the mathematical descrip-tion is different in those components. Although the structure of the biomembrane is described at the molecular level (lipid molecules), the whole approach is at the continuum level.

One should stress for readers with the biological background that an action poten-tial propagates as a travelling pulse in a nerve [6, 25, 34]. In terms of mathematical physics, it is a wave and that is why this Chapter is devoted to the description of waves to demonstrate the fundamental ideas and solutions to wave equations. This is a preparatory step to further interdisciplinary analysis within the framework: physics, mathematics, physiology.

In what follows, the basic physical considerations for dynamical processes are de-scribed. The word 'dynamics' itself comes from Greek: 'dynamis' meaning 'power' and 'dynamikos' meaning 'powerful'. The general description for a wave is the fol-lowing: a wave is a disturbance that travels through the medium (material, tissue); wave motion transfers energy from one point of space to another without transporting matter.

The cornerstone for describing the wave motion is the Newton's second law (cited after Principia translated by Motte [35]) for a moving body:

> The alteration of motion is ever proportional to the motive force impressed; and is made in the direction of the right line in which that force is impressed.

J. Engelbrecht et al., *Modelling of Complex Signals in Nerves*,
https://doi.org/10.1007/978-3-030-75039-8_3

In contemporary wording, this law is expressed as "force is equal to the change of momentum (mv) per change in time". If the mass is constant then we have simply "force equals mass times acceleration":

$$F = ma,\qquad(3.1)$$

where F is the force, m is the mass, a is the acceleration while v is the velocity. In the continuum theory, this law is called balance of momentum [17]. Using the rectangular coordinates, in general three-dimensional case it can be written in terms of stress like [4]

$$(K_{ij}x_{k,j})_{,i} + \rho_0(f_k - A_k) = 0,\qquad(3.2)$$

where K_{ij} is the Kirchhoff stress tensor, x_k are space variables, ρ_0 is the density of the material, f_k are the components of the body force and A_k are the components of the acceleration. Here $k, i, j = 1, 2, 3$ and comma denotes the partial differentiation. The summation over repeated indices is used. The stress is a force per unit area and at a certain point of a body it needs nine components to be specified – shortly expressed by a stress tensor like K_{ij}. Equation (3.2) must be accompanied by the stress-strain relation

$$K_{ij} = K_{ij}(E_{ij}), \quad E_{ij} = \frac{1}{2}\left(u_{i,j} + u_{j,i} + u_{k,i}u_{k,j}\right),\qquad(3.3)$$

where E_{ij} is the Green deformation tensor involving the gradients of displacement u_i.

Like stress tensor, the deformation tensor has nine components. In a one-dimensional case, K_{ij} and E_{ij} are certainly simpler. In the simplest linear one-dimensional case Eq. (3.2) involving the displacement $u_1 = u$, the linear conventional stress-strain relation and neglecting the body force reads:

$$(\lambda + 2\mu)u_{,11} - \rho_0 u_{,tt} = 0,\qquad(3.4)$$

where λ and μ are Lamé coefficients reflecting the elasticity properties of the medium. The comma denotes differentiation with respect to space variable x_1 and time t. This is the simplest wave equation which needs modification to reflect more realistic situations (see, for example, [13]). It is easy to see that $(\lambda + 2\mu)/\rho_0 = c^2$, where c is the velocity.

The balance of momentum is one of the conservation laws, the others are the conservation of mass, the balance of moment of momentum, the conservation of energy (the first principle of thermodynamics) and in addition, the entropy inequality [18]. Further, we shall mostly use the balance of momentum.

It is also possible to derive a wave equation starting from a discrete system of particles which reflects the microscopic structure of materials. For example, the Born-Karman model [29] describes in the one-dimensional setting an infinite elastic chain with particles with the mass m, equidistant by the lattice spacing h and coupled by identical springs of stiffness k_0. Then the longitudinal motion for a particle n is described by Eq. (3.1)

$$m\frac{d^2U_n}{dt^2} = k_0(\xi_{n+1} - \xi_n), \quad \xi_n = U_n - U_{n-1}, \tag{3.5}$$

where U_n is the displacement of a particle. By using the Taylor-series expansion Eq. (3.5) can be transformed in the first approximation to the wave equation similar to Eq. (3.4)

$$U_{,tt} - \frac{k_0 h^2}{m} U_{,11} = 0, \tag{3.6}$$

where the label n is dropped. Here $k_0 h^2/m = c^2$. In both cases - continuum and discrete - the starting point for the derivation of the wave equation is the Newton's second law.

Another important physical process which accompanies the signal propagation in nerve fibres, is thermal conduction. Whatever is the material, this process is governed by the Fourier's law. In the differential form, it is described by

$$\mathbf{Q} = -\alpha \nabla T, \tag{3.7}$$

where vectorial quantity $\mathbf{Q}$ is the local heat flux density, α is the thermal conductivity and ∇T is the local temperature gradient. In the one-dimensional case

$$Q = -\alpha T_{,1}, \tag{3.8}$$

where, as before, the comma denotes the differentiation, this time with respect of x_1.

Clearly, the physical laws are to be followed when modelling the biological processes. However, one should be careful with applications. For example, Cartwright [7] indicates the importance to understand the explanatory nature of laws of physics. She takes the universal law of gravitation (describing force between two bodies) and the Coulomb's law (describing force between two charges) and argues about the principle of *ceteris paribus*. Indeed, both laws, taken separately, describe just one effect which is not justified for charged bodies. She concludes that one should use the explanation by the composition of causes. It seems that this approach is useful in modelling the effects of nerve pulse propagation.

Second, one cannot forget the warning by Fox [20] who mentioned that

... the laws of physics place tough constraints on our mental faculties as well,

meaning possible physical limits in neurophysiology.

3.2 Mathematical Physics

Mathematical physics deals with the development of mathematical methods for solving physical problems. It is very well elaborated for the rigorous analysis of special types of equations. Although the real physical problems are usually described by more sophisticated equations, the knowledge about the basic cases is needed for understanding the backbone of the physical processes. Later for describing the processes in nerve fibres, we rely upon these basic considerations.

Many physical processes are described by a second-order partial differential equation with two independent variables x and y. The corresponding general form is [28, 45]

$$A\Phi_{xx} + 2B\Phi_{xy} + C\Phi_{yy} = H(x, y, \Phi, \Phi_x, \Phi_y), \tag{3.9}$$

where Φ is a vector of dependent variables, A, B, C are constants and $H(\ldots)$ is a function of variables and their first order derivatives. Here and further, the indices denote derivative with respect to indicated independent variables. The three main types of equations are:

(i) hyperbolic equations

$$\frac{1}{c^2}\Phi_{tt} = \nabla^2\Phi; \tag{3.10}$$

(ii) parabolic equations

$$\Phi_t = \alpha\nabla^2\Phi; \tag{3.11}$$

(iii) elliptic equations

$$\nabla^2\Phi = 0. \tag{3.12}$$

Here t is time and ∇^2 denotes the Laplacian

$$\nabla^2\Phi = \Phi_{xx} + \Phi_{yy} + \Phi_{zz}. \tag{3.13}$$

where x, y, z are space coordinates. Hyperbolic and parabolic (diffusion) equations involve time t and describe the dynamical processes under certain initial and boundary conditions. Elliptic equations describe the static state (steady-state) determined by boundary conditions. In our analysis, we pay attention to one-dimensional hyperbolic and parabolic equations.

It is important to note that the hyperbolic equations describe processes with finite velocity c and the parabolic equations – the diffusion processes. The paradox is that the process described by a parabolic equation has an infinite velocity. So seemingly every point in a material 'feels' the heat flux instantaneously. This paradox has been explained by Müller [30] who showed that by extending the theory of thermodynamics it is possible to remove this paradox. Berezovski and Ván [3] have elaborated this theory using the concept of internal variables.

Solutions to the hyperbolic (wave) equation [23].

Let us denote by u the dependent variable, while x and t are the space and time coordinates. As before, the indices denote differentiation. Equation

$$u_{tt} - c^2 u_{xx} = 0, \quad -\infty < x < \infty, \quad t > 0 \tag{3.14}$$

is to be solved under homogenous boundary conditions

$$u(0,t) = 0, \quad u(l,t) = 0, \tag{3.15}$$

where l is the length of the interval. The initial value problem for Eq. (3.14) must satisfy the conditions at $t = t_0$

$$u(x,t_0) = \varphi(x), \quad u_t(x,t_0) = \psi(x), \tag{3.16}$$

where t_0 is the initial moment (usually $t_0 = 0$) and $\varphi(x), \psi(x)$ are smooth functions of their arguments. The D'Alembert solution to the initial value problem is

$$u(x,t) = \frac{1}{2}\left[\varphi(x+ct) + \varphi(x-ct)\right] + \frac{1}{2c}\int_{x-ct}^{x+ct} \psi(x)dx. \tag{3.17}$$

Note that the solution describes two waves – one propagating to the right and another to the left. The uniqueness of this solution is proved, for example, by Tikhonov and Samarski [45].

It is possible to find a solution to Eq. (3.14) in terms of series. Let the initial conditions (3.16) be determined in the interval $0 < x < L$. Then [23]

$$u(x,t) = \sum_{n=1}^{\infty}\left[a_n \cos\left(\frac{n\pi ct}{L}\right) + b_n \sin\left(\frac{n\pi ct}{L}\right)\right]\sin\left(\frac{n\pi x}{L}\right), \tag{3.18}$$

$$a_n = \frac{2}{L}\int_0^L \varphi(x)\sin\left(\frac{n\pi x}{L}\right)dx, \tag{3.19}$$

$$b_n = \frac{2}{n\pi L}\int_0^L \psi(x)\sin\left(\frac{n\pi x}{L}\right)dx. \tag{3.20}$$

For a non-homogenous equation

$$u_{tt} - c^2 u_{xx} = f(x,t), \quad t > 0, \quad 0 < x < L \tag{3.21}$$

subject to homogenous boundary conditions and initial conditions (3.16), the solution is [23]:

$$u(x,t) = \sum_{n=1}^{\infty} U_n(t)\sin\left(\frac{n\pi x}{L}\right) +$$

$$+ \sum_{n=1}^{\infty}\left[a_n \cos\left(\frac{n\pi ct}{L}\right) + b_n \sin\left(\frac{n\pi ct}{L}\right)\right]\sin\left(\frac{n\pi x}{L}\right), \tag{3.22}$$

$$U_n = \frac{2}{n\pi c}\int_0^t \sin\left[\frac{n\pi c(t-\tau)}{L}\right]d\tau\int_0^L f(x,t)\sin\left(\frac{n\pi x}{L}\right)dx, \tag{3.23}$$

$$f(x,t) = \sum_{n=1}^{\infty} f_n(t) \sin\left(\frac{n\pi x}{L}\right), \tag{3.24}$$

$$f_n(t) = \frac{2}{L} \int_0^L f(x,t) \sin\left(\frac{n\pi x}{L}\right) dx, \tag{3.25}$$

with a_n and b_n given by expressions (3.19) and (3.20), respectively.

Solution to parabolic (diffusion) equation [23].

Equation

$$u_t - \alpha^2 u_{xx} = 0, \quad -\infty < x < L, \quad t > 0, \tag{3.26}$$

is to be solved under initial condition

$$u(x,0) = \varphi(x) \tag{3.27}$$

and the homogenous Dirichlet boundary conditions

$$u(0,t) = u(L,t) = 0. \tag{3.28}$$

The corresponding solution is

$$u(x,t) = \sum_{n=1}^{\infty} c_n \exp\left[-\left(\frac{n\pi\alpha}{L}\right)^2 t\right] \sin\left(\frac{n\pi x}{L}\right), \tag{3.29}$$

$$c_n = \frac{2}{L} \int_0^L \varphi(x) \sin\left(\frac{n\pi x}{L}\right) dx. \tag{3.30}$$

Uniqueness of this solution is proved, for example, by Tikhonov and Samarski [45].
Solution for a non-homogenous equation

$$u_t - \alpha^2 u_{xx} = f(x,t), \quad t > 0, \quad 0 < x < L \tag{3.31}$$

subject to zero boundary and initial conditions

$$u(0,t) = u(L,t) = 0, \quad u(x,0) = 0, \tag{3.32}$$

but with existence of external force $f(x,t) \neq 0$. The solution of Eq. (3.31) is [23]

$$u(x,t) = \sum_{n=1}^{\infty} \left\{ \int_0^t \exp\left[-\left(\frac{n\pi\alpha}{L}\right)^2 (t-\tau)\right] f_n(t) d\tau \right\} \sin\left(\frac{n\pi x}{l}\right), \tag{3.33}$$

with

$$f_n(t) = \frac{2}{L} \int\limits_0^L f(x,t) \sin\left(\frac{n\pi x}{L}\right) dx. \tag{3.34}$$

Despite the existence of these analytic solutions, the modified equations are more complicated and solved numerically.

3.3 Wave Equations

The classical wave equation described in Sect. 3.2 describes the waves in the homogeneous medium, it is a linear equation and losses are not accounted for. Beautiful and simple as this equation is, it must be modified to describe reality better and capture the accompanying effects. In mathematical terms, it means that in addition to the second-order derivatives, the higher or also the lower derivatives of the dependent variable with respect to space and time will appear in the governing equation. In what follows, the possible modifications of the classical wave equation are presented describing nonlinear, dissipative, dispersive and thermal effects. A more detailed discussion on such modified mathematical models is presented by Engelbrecht [13].

The one-dimensional setting is used to make the presentation as simple as possible. The dependent variable is denoted by u, space and time coordinates are denoted by x and t, respectively, while indices x and t denote the differentiation. Implicitly it is assumed that the changes introduced by modifications are of the order larger than just small quantitative corrections but may lead to qualitative changes in wave profiles (see further Sect. 3.5). One should also note the axiom of equipresence which says that all the effects of the same order should be taken into account simultaneously [13, 18].

Let us start from the derivation of the wave equation based on Eqs. (3.2)-(3.3). For the sake of simplicity we take $f_1 = 0$. The deformation tensor is

$$E_{11} = \left(u_x + \frac{1}{2}u_x^2\right). \tag{3.35}$$

The stress tensor K_{ij} is related to the Helmholtz free energy F by [4]

$$K_{ij} = \rho_0 \frac{\partial F}{\partial E_{ij}}. \tag{3.36}$$

If we use only a quadratic formulation of the free energy then

$$K_{11} = (\lambda + 2\mu)u_x \tag{3.37}$$

and the wave equation is like Eq. (3.4):

$$u_{tt} - c_0^2 u_{xx} = 0, \quad c_0^2 = \frac{\lambda + 2\mu}{\rho_0}. \tag{3.38}$$

If the free energy is of the cubic form then additional (third order) elastic constants ν_1, ν_2, ν_3 appear [4] and the stress is

$$K_{11} = (\lambda + 2\mu)u_x + \left(\frac{1}{2}\lambda + \mu + 3\nu_1 + 3\nu_2 + 3\nu_3\right)u_x^2. \tag{3.39}$$

Then the resulting wave equation is

$$u_{tt} - c_0^2\left[1 + 3(1 + m_0)u_x\right]u_{xx} = 0, \tag{3.40}$$

where $m_0 = 2(\nu_1 + \nu_2 + \nu_3)/(\lambda + 2\mu)$. The coefficient m_0 reflects the coupled influence of geometrical (due to the deformation tensor) and physical (due to stress tensor) nonlinearities (more details given in [13]). Note that in most conventional theories of wave motion in solids [4, 13, etc.] the nonlinearities are of the gradient-type, i.e., dependent on u_x.

If the stress tensor K_{ij} includes also the irreversible part (time-dependent) due to the viscosity of the material, then the governing equation takes the form [8]

$$u_{tt} - c_0^2 u_{xx} - \nu u_{xxt} = 0, \tag{3.41}$$

where ν is the coefficient of viscosity. This is the third order partial differential equation, usually called the Voigt model [8].

If thermal effects are taken into account, then the conventional theory of thermoelasticity in terms of u and temperature Θ is used. Then the governing equations are [36]:

$$u_{tt} - c_0^2 u_{xx} - \kappa\rho_0^{-1}\Theta_x = 0, \tag{3.42}$$

$$\rho_0 c_p \Theta_t = \alpha\Theta_{xx} + m\Theta_0 u_{xt}, \tag{3.43}$$

where Θ_0 is the initial temperature and the coefficients κ, c_p, α, m are related to the thermal properties of a material [13, 36]. In this case the governing equations involve hyperbolic and parabolic equations (see Sect. 3.2). More sophisticated models are described by Berezovski and Ván [3].

Waves are often influenced by dispersive effects. Dispersion means that the waves of various frequencies (various wavelengths) propagate with different velocities. The sources for dispersion can be geometrical (the existence of a free surface like in rods) or physical (the existence of a microstructure). In both cases, the higher-order derivatives appear in the wave equation.

The linear waves in Mindlin-type microstructured solids [13, 37] are described by the following equation:

$$u_{tt} - (c_0^2 - c_A^2)u_{xx} = p^2 c_1^2(u_{tt} - c_0^2 u_{xx})_{xx} - p^2(u_{tt} - c_0^2 u_{xx})_{tt}. \tag{3.44}$$

An asymptotic version of this equation is

$$u_{tt} - (c_0^2 - c_A^2)u_{xx} = p^2 c_A^2(u_{tt} - c_1^2 u_{xx})_{xx}. \tag{3.45}$$

Here c_A and c_1 are velocities which characterise the coupling effects and microstructure, respectively and p is a certain scale factor. In Eq. (3.45), the wave operator on the left describes macrostructure and the wave operator on the right – the microstructure. It is obvious that for the microstructure the inertial (term u_{ttxx}) and elastic (term u_{xxxx}) effects are taken into account. Both terms are needed for grasping dispersive effects [37, 44]. Further we call Eq. (3.44) the full equation and Eq. (3.45) – the hierarchical equation. In the nonlinear case (c.f. Eq. (3.40) with nonlinearities at both macro- and microscale, the equation is [44]:

$$u_{tt} - b^2 u_{xx} - \frac{1}{2}\mu\left(u_x^2\right)_x = \delta\left[\beta u_{tt} - \gamma u_{xx} - \frac{1}{2}\lambda\delta^{1/2}\left(u_{xx}^2\right)_x\right]_{xx}, \qquad (3.46)$$

where $b, \beta, \gamma, \mu, \lambda$ are physical parameters and $\delta = l/L$ is the scale parameter. Here l is the scale of the microstructure and L is the wavelength.

The wave equations involving nonlinearity and dispersion are called Boussinesq-type equations [9]. More sophisticated models for double-scale microstructure (scale within a scale) include even higher-order derivatives [13]. The governing equations for longitudinal waves in rods are similar to those in microstructured solids including different parameters reflecting also the geometrical properties of rods and the Rayleigh correction relating the transverse displacement to longitudinal deformation [39, 41].

Finally, let us turn our attention to electric waves in conductors. By using Ohm's law and the balance of flows, the so-called telegraph equations can be derived [28]:

$$v_x + Li_t + Ri = 0, \qquad (3.47)$$

$$i_x + Cv_t + Gv = 0. \qquad (3.48)$$

Here i is the current and v – the potential (voltage) while L denotes the self-inductance, R – the resistance, C – the capacitance and G – the leakage (losses). It is possible to write the system of two first-order equations in one second-order equation (telegraph equation)

$$v_{xx} = LCv_{tt} + (RC + GL)v_t + RGv. \qquad (3.49)$$

Introducing the new variable

$$v = \exp\left(-\frac{b}{a}t\right)u, \qquad (3.50)$$

Eq. (3.49) is transformed to

$$u_{tt} - c_0^2 u_{xx} = b_0^2 u, \qquad (3.51)$$

where $c_0^2 = 1/a$, $v_0^2 = (b^2 - ac)/a$ and $a = LC, 2b = RC + GL, c = GR$.

Equation (3.51) is a wave equation with an additional term involving the leakage. This equation has an analytical solution in terms of the Bessel function of the zero-order [28].

3.4 Reaction-Diffusion Equations

If diffusion is related to possible production or consumption of a modelled variable then the outcome is a reaction-diffusion equation. Compared with the classical parabolic (diffusion) equation a source term is added and the general form of such an equation is (c.f. Eq. (3.11))

$$\alpha \Phi_t = \nabla^2 \Phi + R(\Phi), \tag{3.52}$$

where $R(\Phi)$ is the vector of local reactions. In the simple one-dimensional case one has

$$u_t - \alpha u_{xx} = f(u), \tag{3.53}$$

where α is a coefficient. Fisher [19] has proposed $f(u) = ru(1 - u), r = const.$ for describing the dynamics of population. Equation (3.52) is also called the Kolmogorov-Petrovsky-Piskunov equation [26]. This equation is a cornerstone of reaction-diffusion equations. Many other types of $f(u)$ are studied, for example, $f(u) = u(1 - u^2)$ describes the Rayleigh-Bénard convection, $f(u) = au(1 - u/K)$, $a = const., K = const.$ describes logistic growth, etc. If diffusion is neglected, then

$$\dot{u} = f(u) \tag{3.54}$$

is an ordinary differential equation. The two-dimensional cases of Eq. (3.52) are extremely interesting and as mentioned by Scott [42] lead us "into as garden of unexpectedly beautiful phenomena" because of emerging patterns due to the self-organisation.

Another basic example of a reaction-diffusion system is the Lotka-Volterra model which describes the predator-prey dynamics [31]. Its diffusive form is

$$u_t - D_1 u_{xx} = u(1 - v), \tag{3.55}$$

$$v_t - D_2 v_{xx} = av(u - 1), \tag{3.56}$$

where D_1, D_2 are diffusion coefficients and $a = const.$ The most studied form neglects diffusion and then we have

$$\dot{u} = u(1 - v), \tag{3.57}$$

$$\dot{v} = av(u - 1). \tag{3.58}$$

The Lotka-Volterra model and its modifications have been widely used not only to predict predator-prey dynamics but also for modelling dynamics of cells and infectious deceases. Especially interesting is the idea of cross-diffusion models with diffusion of both variables [24]

$$u_t - d_1 u_{xx} - d_2 v_{xx} = u f(x, u, v), \qquad (3.59)$$

$$v_t - d_3 v_{xx} - d_4 u_{xx} = v g(x, u, v), \qquad (3.60)$$

where d_1, d_2, d_3, d_4 are diffusion constants and $f(\ldots), g(\ldots)$ – smooth functions.

Let us return to the telegraph equation (Sect. 3.3). If we neglect inductivity and add additional current j_i to the balance then Eqs. (3.47) and (3.48) yield

$$v_x + Ri = 0, \qquad (3.61)$$

$$i_x + C v_t + j_i = 0. \qquad (3.62)$$

Combining Eqs. (3.61) and (3.62) we obtain

$$RC v_t - v_{xx} + R j_i = 0. \qquad (3.63)$$

This is the basis for the Hodgkin and Huxley [22] for deriving their celebrated model for the action potential. This model will be presented later in more detail.

The simplified FitzHugh-Nagumo model has also the similar structure. In original notations [32] it reads:

$$h u_{ss} = \frac{1}{c} u_t - w - \left(u - \frac{u^3}{3} \right), \qquad (3.64)$$

$$c w_t + b w = a - u, \qquad (3.65)$$

where a, b, c, h are constants and $s \equiv x$.

3.5 Physical Effects

The mathematical models described above can describe definite physical effects which reflect the properties of the medium. According to the classical wave equation (3.14) the generated wave propagates without any changes. The modified wave equations describe more complicated situations and the profiles of waves may undergo essential changes during the propagations. In this process, the intensity of waves (amplitude or energy) also plays an important role. Many studies describe such physical effects [5, 46, etc.]. These effects demonstrate the richness of the physical world and the power of mathematics for describing them. Metaphorically speaking [10], "mathematics is biology's microscope for discovering physical effects."

In what follows, the main physical effects that may occur important for the further analysis of nerve signals are briefly described. This description is mainly based on the earlier studies of authors [2, 13, 16, etc.].

Nonlinearity.

The simplest nonlinear wave equation is Eq. (3.40). Let us rewrite it in the form

$$u_{tt} - c^2 u_{xx} = 0, \tag{3.66}$$

where $c^2 = c_0^2 \, [1 + 3(1 + m_0)u_x]$. This means that the velocity depends on the gradient u_x. Equation (3.66) is still hyperbolic but its characteristics have various slopes due to changes in the velocity and may intersect. In physical terms it means that a shock wave may emerge, in mathematical terms it means a discontinuity (a jump). A scheme of such a process is shown in Fig. 3.1.

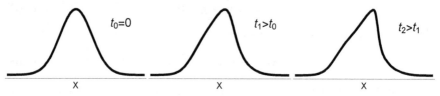

Fig. 3.1 Scheme of emerging a shock wave.

In this process, one should understand the importance of the material properties reflected by $3(1 + m_0)$. For usual materials $m_0 < 0$ and is of the order of 10 [12]. It means that for such material, a compressive shock wave may emerge [5]. If, however, $m_0 > 0$ then tensile shock wave may emerge. The emerging of shock waves is intensively studied in nonlinear acoustics [33].

Dispersion.

In simple terms, dispersion means the dependence of wave velocity on frequency (or wavelength). Dispersion analysis is based on studying the properties of wave described by [46]:

$$u(x,t) = \hat{u} \exp[i(kx - \omega t)], \tag{3.67}$$

where k is the wave number, ω - the frequency and $\hat{u}$ - the amplitude. One should distinguish the phase c_{ph} and group c_{gr} velocities:

$$c_{ph} = \frac{\omega}{k}, \quad c_{gr} = \frac{d\omega}{dk}. \tag{3.68}$$

The phase velocity is the velocity of any component of a wave. In simple terms, it is a velocity of a crest. The group velocity describes the velocity of the overall envelope and is often understood as the velocity at which energy or information propagates.

For the classical wave equation (3.14)

$$\omega = ck \tag{3.69}$$

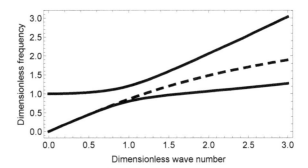

Fig. 3.2 Dispersion curves for models (3.44) and (3.45). Dispersion relation (3.70) is plotted by solid and (3.71) by dashed line. Here $c_A/c_0 = 0.4$ and $c_1/c_0 = 0.3$.

which means that $c_{ph} = c_{gr} = c$ and there is no dispersion. In more complicated cases dispersion relation $\omega = \omega(k)$ gives a lot of information on the process. In case of the microstructured medium, the governing equation (3.44) yields

$$\omega^2 = (c_0^2 - c_A^2)k^2 + p^2(\omega^2 - c_0^2 k^2)(\omega^2 - c_1^2 k^2) \qquad (3.70)$$

and for the governing equation (3.45) results

$$\omega^2 = (c_0^2 - c_A^2)k^2 - p^2 c_A^2 (\omega^2 - c_1^2 k^2)k^2. \qquad (3.71)$$

Fig. 3.3 Group (top panel) and phase (bottom panel) velocity curves for models (3.44) and (3.45). Velocity curves corresponding to dispersion relation (3.70) are represented by solid lines; while dashed lines represent dispersion relation (3.71). Here $c_A/c_0 = 0.4$ and $c_1/c_0 = 0.3$.

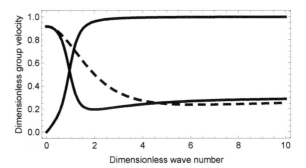

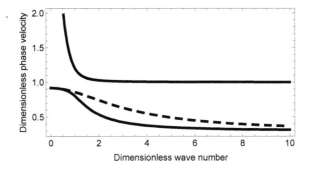

Fig. 3.4 Top panel: wave
profiles of model (3.44) (solid
line) and (3.45) (dashed line)
for an impulse type boundary
condition at 50 time steps.
Bottom panel: corresponding
group velocity curves. Here
$c_A/c_0 = 0.7$ and $c_1/c_0 = 0.3$.

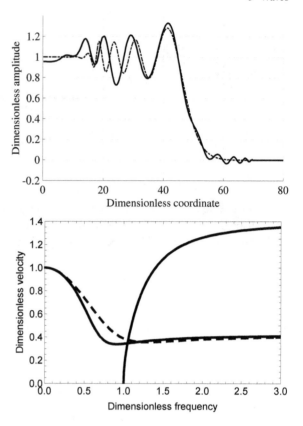

Dispersion relations (3.70) and (3.71) are shown in Fig. 3.2 where dimensionless
frequency $\eta = p\omega$ and dimensionless wave number $\xi = c_0 p k$ have been introduced
for convenience. Dispersion relation (3.70) is represented by solid lines and consists
of two branches. Traditionally the lower branch is called the acoustic and higher
branch the optical branch. Dispersion relation (3.71) is represented by dashed line.
Alternatively, dispersion can be visualised in terms of phase and group velocity
curves (see Eq. (3.68)). This is demonstrated in Fig. 3.3. The velocity curves can
also be presented in terms of frequency against velocity.

Model (3.45) is a hierarchic approximation for model (3.44). It is derived by
making use of the slaving principle [14] and is reflected by dispersion curves in the
absence of higher frequency optical branch (see Figs. 3.2 and 3.3).

The effect of dispersion on the wave propagation is demonstrated in Fig. 3.4 where
models (3.44) and (3.45) are solved under impulse boundary condition making use
of the Laplace transform (see [38] for details). The effect of the optical branch is
seen in the oscillating tail in front of the main pulse, which is caused by the higher
frequency harmonics travelling at higher velocities (see Fig. 3.4 bottom).

Nonlinearity and dispersion.

The nonlinear effects lead to steepening of the wave profile up to the discontinuity while dispersive effects tend to scatter the profile. If these effects are balanced then an interesting effect will emerge. Namely, as a result of such a balance, the steady wave profiles may emerge called solitons. In our presentation, Eq. (3.46) is a governing equation for emerging solitons.

Soliton is by definition a nonlinear wave which maintains its shape, propagates with a constant velocity and restores its shape and velocity after collision with another soliton, except the phase shift (see, for example, Ablowitz [1]). The soliton theory started after John Scott Russell described in 1834 such a wave in a canal and Korteweg and de Vries [27] have derived an equation for describing such a phenomenon. This equation

$$v_t + v v_\xi + v_{\xi\xi\xi} = 0, \tag{3.72}$$

where v is an amplitude and τ, ξ are moving coordinates is called the Korteweg-deVries (KdV) equation. Note that the KdV equation is an one-wave equation contrary to the two-wave Boussinesq equations (like Eq. (3.46)).

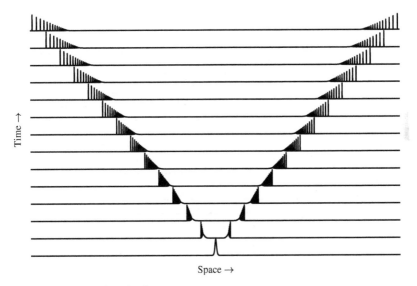

Fig. 3.5 Formation of trains of solitons.

In order to demonstrate the existence of solitons and formation of soliton trains, Eq. (3.46) is integrated numerically using the pseudospectral method [40]. Given the initial condition at $X = X_0$

$$U(X,0) = U_0 \operatorname{sech}^2 B_o (X - X_0) \tag{3.73}$$

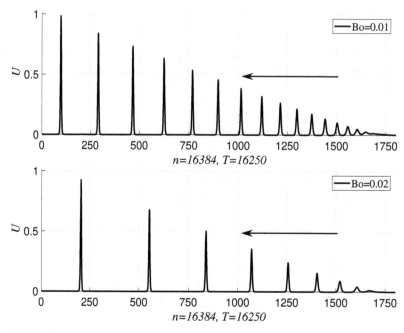

Fig. 3.6 Single wave profiles at $T = 16250$ for Eq. (3.46). U is the amplitude.

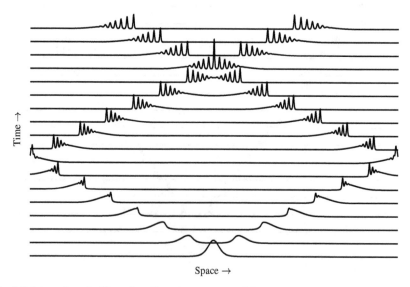

Fig. 3.7 Interaction of solitons for a Boussinesq-type model.

and periodic boundary condition, the emergence of soliton trains can be analysed numerically (see Fig. 3.5).

The number of solitons in the train depends on the width of the initial pulse B_o. This effect can be seen in Fig. 3.6 for two values of B_o.

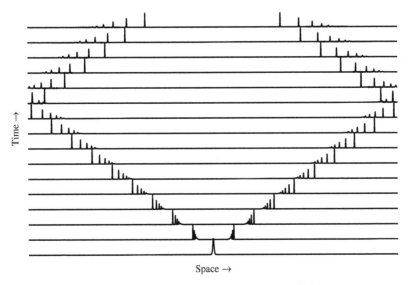

Fig. 3.8 Interaction of solitary waves for Eq. (3.46) demonstrating radiation.

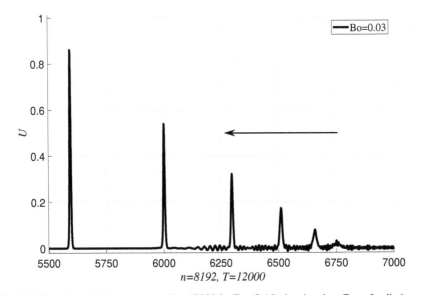

Fig. 3.9 The train of solitary waves at $T = 12000$ for Eq. (3.46) showing the effect of radiation.

Interaction of solitons is demonstrated in Fig. 3.7 where it can be seen that upon collision soliton amplitude is greater than the amplitudes of interacting solitons meaning that the interaction of solitons is nonlinear. This example corresponds to solitons in a biomembrane [15].

The question is whether the emerged solitons depicted in Figs. 3.5 and 3.6 are really solitons, i.e., whether their interaction is elastic like in case of processes modelled by the KdV equation. The numerical calculations demonstrate that the radiation due to the interactions starts to influence mostly the solitons with small amplitude. This is demonstrated in Figs. 3.8 and 3.9 for Eq. (3.46). Therefore, solitary waves modelled by the Boussinesq-type equations are not solitons in a strict sense.

Dissipation.

In general terms dissipation means that the amplitude of waves decreases during the propagation. The reason of such an irreversible process may be related to the internal friction like in the Voigt model (see Eq. (3.41)) or to thermal losses (see Eqs. (3.42) and (3.43)). We are not going into details and refer here to monographs by Christensen [8] and Nowacki [36]. Figure 3.10 depicts schematically the effects of dissipation.

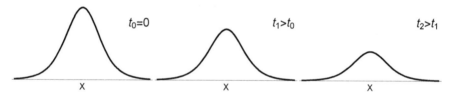

Fig. 3.10 Evolution of a travelling pulse under dissipation.

3.6 The Wave Equation with Forcing

It is important to understand the effect of external forcing in wave motion. In Sect. 3.2 we have presented an analytical solution to the nonhomogenous equation (3.21). Further, we need such solutions for more complicated equations and then the numerical simulation is used. Here we present preliminary results of solving the forced wave equation under special types of external forces: unipolar and bipolar cases. These forces are shown in Fig. 3.11.

Classical wave equation (3.14) and its solutions are discussed in Sect. 3.2. Homogenous wave equation

$$u_{tt} - c^2 u_{xx} = 0 \tag{3.74}$$

Fig. 3.11 Two types of external forcing – unipolar (top panel) and bipolar (bottom panel).

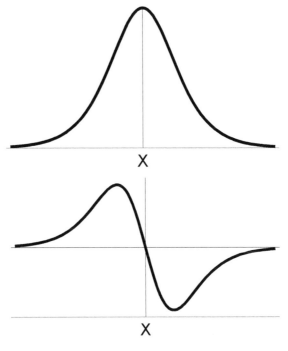

Fig. 3.12 Wave profile under moving bipolar external forcing. Solid black line represent moving bipolar forcing and dashed blue line represent emerged wave profile under this forcing.

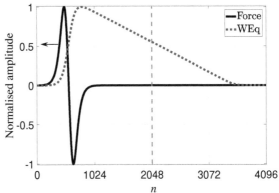

describes a travelling wave with constant shape. In case of inhomogeneous wave equation

$$u_{tt} - c^2 u_{xx} = f(x,t) \tag{3.75}$$

the shape of the wave profiles in influenced by the external forcing. In order to demonstrate this a following equation is solved with pseudospectral method:

$$v_{tt} - c^2 v_{xx} = \gamma u_x \tag{3.76}$$

with $u(x, 0) = \text{sech}^2(x)$ and $v(x, 0) = 0$. This means that γu_x acts like bipolar forcing (see Fig. 3.11).

Figure 3.12 shows a wave profile emerging under bipolar forcing (blue dotted line). Black solid line represents the bipolar forcing and the vertical red dashed line shows the initial position. While bipolar forcing only travels to left, the emerging wave profile travels to both directions.

3.7 Solitary Waves and solitons

Solitary waves are often met in various physical problems. Sometimes a question is raised: are all solitary waves solitons? Here we may ask this question with regard of the action potential (AP) that is why we should explain the situation in more detail. Typical soliton and AP profiles are shown in Fig. 3.13.

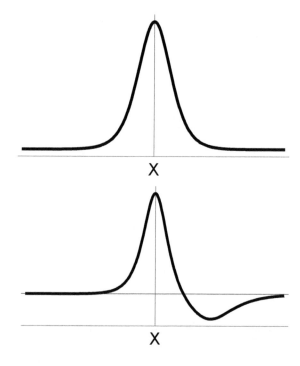

Fig. 3.13 Schematic profiles of a soliton (top panel) and an action potential (bottom panel).

Let us repeat a definition for a soliton (see Sect. 3.5): soliton is a solitary wave which maintains its shape, propagates with a constant velocity and restores its shape and velocity after collision with another soliton except the phase shift [1]. The action potential is an asymmetric solitary pulse with an overshoot. Here and further, we consider soliton as the classical solution of the KdV equation (3.72) and the action potential as a solution of a reaction-diffusion system. Certainly, both entities are also experimentally observed. The notion of soliton is coined by Zabuski and Kruskal [47] who showed that the famous Fermi-Pasta-Ulam problem about the thermal equilibrium in a nonlinear lattice was in the continuum limit described by the KdV equation. They noticed the resemblance of the numerically obtained solitary wave with particles (c.f. electron, photon, etc.) that is why they called such a solution by the name 'soliton'.

There are essential differences between a soliton and the AP [42]. First, the velocity of a soliton depends on its amplitude while the amplitude of an AP depends on the properties of the medium. Second, a collision between two solitons retains their shapes and velocities except the phase shift but two colliding APs annihilate each other. Third, there is a threshold condition for generating an AP and due to an overshoot, the next AP can be generated only after the relaxation time while the generation of a soliton from an arbitrary input depends on the energy of this input. It is possible that from one input a train of solitons will emerge. One should note, however, that the solitons described by the Boussinesq-type equations start to lose energy after several collisions due to radiation effects [9].

Describing the difference between solitons and APs, it does not mean that solitons cannot be met in biological systems. Biomembranes, as shown by Heimburg and Jackson [21] can carry solitons. The localised excitations in an α-helix of proteins are called Davydov solitons [11]. Nature is rich with exciting phenomena. Soliton is a cornerstone of contemporary nonlinear wave motion and expresses one of the ideas of complexity theory: the interaction between two competing physical effects leads to a new quality.

References

1. Ablowitz, M.J.: Nonlinear Dispersive Waves. Asymptotic Analysis and Solitons. Cambridge University Press, Cambridge (2011). DOI 10.1017/CBO9780511998324
2. Berezovski, A., Engelbrecht, J., Salupere, A., Tamm, K., Peets, T., Berezovski, M.: Dispersive waves in microstructured solids. Int. J. Solids Struct. **50**(11-12), 1981–1990 (2013)
3. Berezovski, A., Ván, P.: Internal Variables in Thermoelasticity, Springer, Cham (2017). DOI 10.1007/978-3-319-56934-5
4. Bland, D.R.: Nonlinear Dynamic Elasticity. Blaisdell Publishing Company, Waltham (1969)
5. Bland, D.R.: Wave Theory and Applications. Clarendon Press, Oxford (1988)
6. Bressloff, P.C.: Waves in Neural Media. Springer, New York (2014). DOI 10.1007/978-1-4614-8866-8
7. Cartwright, N.: How the Laws of Physics Lie. Oxford University Press (1983). DOI 10.1093/0198247044.001.0001
8. Christensen, R.: Theory of Viscoelasticity: An Introduction. Elsevier (2012)
9. Christov, C.I., Maugin, G.A., Porubov, A.V.: On Boussinesq's paradigm in nonlinear wave propagation. Comptes Rendus Mécanique **335**(9-10), 521–535 (2007). DOI 10.1016/j.crme.2007.08.006
10. Cohen, J.E.: Mathematics is biology's next microscope, only better; biology is mathematics' next physics, only better. PLoS Biol. **2**(12), e439 (2004). DOI 10.1371/journal.pbio.0020439
11. Davydov, A.: Solitons and energy transfer along protein molecules. J. Theor. Biol. **66**(2), 379–387 (1977). DOI 10.1016/0022-5193(77)90178-3

12. Engelbrecht, J.: Nonlinear Wave Dynamics. Complexity and Simplicity. Kluwer, Dordrecht (1997). DOI 10.1007/978-94-015-8891-1

13. Engelbrecht, J.: Questions About Elastic Waves. Springer International Publishing, Cham (2015). DOI 10.1007/978-3-319-14791-8

14. Engelbrecht, J., Berezovski, A., Pastrone, F., Braun, M.: Waves in microstructured materials and dispersion. Philos. Mag. **85**(33-35), 4127–4141 (2005). DOI 10.1080/14786430500362769

15. Engelbrecht, J., Tamm, K., Peets, T.: On solutions of a Boussinesq-type equation with displacement-dependent nonlinearities: the case of biomembranes. Philos. Mag. **97**(12), 967–987 (2017). DOI 10.1080/14786435.2017.1283070

16. Engelbrecht, J., Salupere, A., Berezovski, A., Peets, T., Tamm, K.: On nonlinear waves in media with complex properties. In: H. Altenbach, J. Pouget, M. Rousseau, B. Collet, T. Michelitsch (eds.) Generalized Models and Non-classical Approaches in Complex Materials 1, *Advanced Structured Materials*, vol. 89, pp. 275–286. Springer, Cham (2018)

17. Eringen, A.C.: Nonlinear Theory of Continuous Media. McGraw-Hill Book Company, New York (1962)

18. Eringen, A.C., Maugin, G.A.: Electrodynamics of Continua I. Springer New York, New York, NY (1990). DOI 10.1007/978-1-4612-3226-1

19. Fisher, R.A.: The wave of advance of advantageous genes. Ann. Eugen. **7**(4), 355–369 (1937). DOI 10.1111/j.1469-1809.1937.tb02153.x

20. Fox, D.: The limits of intelligence. Sci. Am. **305**(1), 36–43 (2011)

21. Heimburg, T., Jackson, A.D.: On soliton propagation in biomembranes and nerves. Proc. Natl. Acad. Sci. USA **102**(28), 9790–5 (2005). DOI 10.1073/pnas.0503823102

22. Hodgkin, A.L., Huxley, A.F.: Resting and action potentials in single nerve fibres. J. Physiol. **104**, 176–195 (1945)

23. Jeffrey, A., Dai, H.H.: Handbook of Mathematical Formulas and Integrals, 4th edn. Academic Press, Amsterdam et al. (2008)

24. Jüngel, A.: Diffusive and nondiffusive population models. In: Mathematical Modeling of Collective Behavior in Socio-Economic and Life Sciences, vol. 51, pp. 397–425. Birkhäuser Boston, Boston (2010)

25. Kaufmann, K.: Action Potentials and Electromechanical Coupling in the Macroscopic Chiral Phospholipid Bilayer. Caruaru, Brazil (1989)

26. Kolmogorov, A., Petrovskii, I., Piskunov, N.: A study of the diffusion equation with increase in the amount of substance. Moscow Univ. Math. Bull. **1**, 1–25 (1937)

27. Korteweg, D.J., de Vries, G.: On the change of form of long waves advancing in a rectangular canal, and on a new type of long stationary waves. Philos. Mag. **39**(240), 422–443 (1895). DOI 10.1080/14786449508620739

28. Koshlyakov, A., Smirnov, M., Gliner, E.: Differential Equations of Mathematical Physics. North-Holland Pub. Co., Amsterdam et al. (1964)

29. Maugin, G.A.: Nonlinear Waves in Elastic Crystals. Oxford University Press, Oxford (1999)

30. Müller, I.: Zum paradoxon der wärmeleitungstheorie. Zeitschrift für Phys. **198**(4), 329–344 (1967). DOI 10.1007/BF01326412

31. Murray, J.: Mathematical Biology: I. An Introduction., 3rd edn. Springer, Berlin (2007)

32. Nagumo, J., Arimoto, S., Yoshizawa, S.: An active pulse transmission line simulating nerve axon. Proc. IRE **50**(10), 2061–2070 (1962). DOI 10.1109/JRPROC.1962.288235

33. Naugolnykh, K.A., Ostrovsky, L.A.: Nonlinear Wave Processes in Acoustics. Cambridge University Press, Cambridge (1998)

34. Nelson, P.C., Radosavljevic, M., Bromberg, S.: Biological Physics: Energy, Information, Life. W.H. Freeman and Company, New York, NY (2003)

35. Newton, I.: The Principia. Translation by A. Motte. Prometheus Books, Amherst (1995)

36. Nowacki, W.: Thermoelasticity, 2nd edn. Pergamon Press, Oxford (1986)

37. Peets, T.: Dispersion Analysis of Wave Motion in Microstructured Solids. Ph.D. thesis, Tallinn University of Technology (2011)

38. Peets, T.: Internal scales and dispersive properties of microstructured materials. Math. Comput. Simul. **127**, 220–228 (2016). DOI 10.1016/j.matcom.2014.03.006

39. Porubov, A.V.: Amplification of Nonlinear Strain Waves in Solids. World Scientific, Singapore (2003)
40. Salupere, A.: The pseudospectral method and discrete spectral analysis. In: E. Quak, T. Soomere (eds.) Applied Wave Mathematics, pp. 301–334. Springer Berlin Heidelberg, Berlin (2009). DOI 10.1007/978-3-642-00585-5
41. Samsonov, A.: Strain Solitons in Solids and How to Construct Them. Chapman and Hall/CRC, Boca Raton (2001)
42. Scott, A.: Nonlinear Science. Emergence and Dynamics of Coherent Structures. Oxford University Press (1999)
43. Stewart, I.: In Pursuit of the Unknown: 17 Equations That Changed the World. Profile Books, London (2013)
44. Tamm, K.: Wave Propagation and Interaction in Mindlin-Type Microstructured Solids: Numerical Simulation. Ph.D. thesis, Tallinn University of Technology (2011)
45. Tikhonov, A., Samarski, A.: Partial Differential Equations of Mathematical Physics. Vol. 1. Holden-Day, University of Michigan (1964)
46. Whitham, G.: Linear and Nonlinear Waves. Wiley, New York (1974)
47. Zabusky, N.J., Kruskal, M.D.: Interaction of "solitons" in a collisionless plasma and the recurrence of initial states. Phys. Rev. Lett. **15**(6), 240–243 (1965). DOI 10.1103/PhysRevLett.15.240

Part II
Dynamical Processes in Nerve Axons

Chapter 4
Nervous Signals

> *The nervous system does its operations in a number of different ways.*
>
> *Laura Bennet, 2000*

The studies on nerve pulse propagation are historically related to the analysis of action potentials which have been measured using electrophysical techniques. The long history briefly described in Chap. 1 (Introduction) is full of remarkable results but the contemporary understanding is based on celebrated studies of A.L. Hodgkin and A.F. Huxley in 50ies-60ies of the 20th century.

Leaving aside the neural networks, the main structural element to be studied is the axon along which an electrical signal (information) propagates from the cell body to the nerve terminal. The morphology of axons is well understood [6]. A simplified scheme of a neuron and an axon is shown in Fig. 4.1

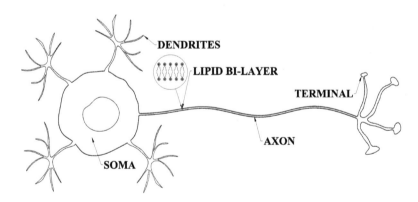

Fig. 4.1 Simplified sketch of a neuron (scales given in the text).

Axons can be myelinated or unmyelinated. In the first case, axons are covered with insulating myelin sheaths interrupted by nodes of Ranvier, in the second case such a myelin sheath is absent. The types of myelinated and unmyelinated axons for mammalians are listed by Debanne et al. [6]. The famous experiments of Hodgkin and Huxley [13] were carried on with the squid giant axons which were unmyelinated. Further in this book, we build up the models for unmyelinated axons.

An axon can be modelled as a tube in a certain environment. Inside the tube is the axoplasmic fluid (shortly axoplasm), called also intracellular fluid. This fluid contains cytoskeletal filaments and has a certain concentration of ions [19]. The wall of the tube has a bilayered lipid structure called biomembrane [19].

A biomembrane is composed of two layers of amphiphilic phospholipids with hydrophilic heads and hydrophobic tails. Such biomembranes are important building blocks of cells in general. A scheme of an axon with its main structural elements is shown in Fig. 4.2.

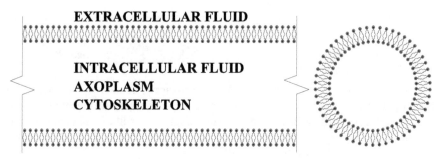

Fig. 4.2 Simplified scheme of axon; longitudinal and transverse (scales given in the text).

The structure of a biomembrane in the longitudinal direction is inhomogeneous because it contains the ion channels composed by certain proteins. These ion channels play the most important role in the propagation of an electrical signal along the axon. Namely, the ion currents, i.e., the flow through these channels regulate the shape of the signal by electrochemical gradients. During the propagation of a signal, the ion concentrations in extra- and intracellular fluids are changing due to these gradients. Channels are mostly voltage-gated but can be also mechanically sensitive [19]. The scheme of an ion channel is shown in Fig. 4.3. Note that the distribution of phospholipids in a membrane is uneven between the two sides of a membrane (see [18]). This effect is not taken into account in our model.

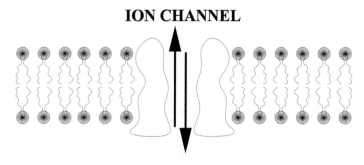

Fig. 4.3 Simplified scheme of an ion channel (scales given in the text).

The dimensions of axons are highly variable [6]. The length may vary between 1 mm to 1 m. The mammalian neurons may have a diameter around 20 μm but in the brain between 0.08 and 0.4 μm. The thickness of the biomembrane (hydrophilic and hydrophobic parts together) is about 5 nm.

Hodgkin and Huxley have experimentally measured the electrical signal – action potential (AP) in a giant squid axon. Their experiment is a fundamental result which has paved the road for further studies and is shown in Fig 4.4.

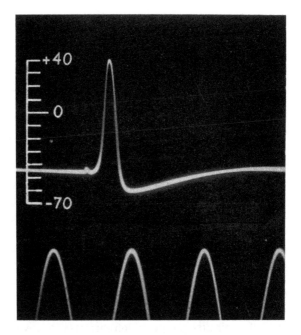

Fig. 4.4 The first intracellular recording of an action potential. Reproduced with permission from [12]; ©Springer Nature 1939.

Schematically such an asymmetric AP is shown in Fig. 4.5.

The AP can be characterised by the following specific features [20]:

- it is an all-or-nothing response: if the stimulus is below threshold then there will be no response far from the stimulating point; if the stimulus is above threshold then a travelling wave is created whose peak amplitude does not depend on the strength of the stimulus;

- the AP propagates along the axon at a constant velocity dictated by the properties of the axon; the changes in velocity are related to the variation of the diameter of the axon [23]; the velocity also depends on the distance between the ion channels;

- the peak of the AP is found independent of distance and preserves its shape; however, later experiments have shown a possible decrease in the amplitude [2], the amplification [6] or the broadening of the AP [25];

- after the passage of an AP, the membrane potential has an overshoot and then slowly recovers;

- before the resting potential is restored, no new AP can be generated which means the existence of a refractory period.

Fig. 4.5 A sketch of a typical
action potential.

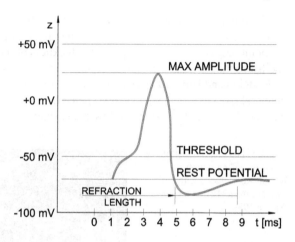

The physical data of the AP measured by Hodgkin and Huxley [13] were (see Fig. 4.4): the largest amplitude measured from the resting state was about 90 mV and the resting potential itself was about ca -45 mV, the full length of the AP including the refraction period was about 4 m s^{-1}. The velocity of the AP measured was 18.8 m s^{-1}.

According to Hodgkin and Huxley [13], the ion currents have the main role in the generation of an AP. In their model, the potential triggers the membrane depolarisation, the inward sodium turn-on, the outward potassium turn-on and the sodium turn-off are decisive factors in the process. These flows mean the movement of ions through the membrane via voltage-gated ion channels. Ion currents are assumed to behave independently and are described by the corresponding phenomenological variables governed by corresponding kinetic equations (see Chap. 5). This allows also calculating the profiles of ion currents in time with respect to the AP [14] which could also explain the length of the refractory period.

The electrophysiology has been developed through the last half a century since Hodgkin and Huxley developed their model which has become a paradigm in neural sciences. The recent understandings are described in many studies [4, 5, 6, etc]. It has been found that not only the sodium and potassium ions play role in the AP formation but also many other ions, like Ca^{2+}, Cl^-, etc. However, the ionic mechanisms like described by Hodgkin and Huxley form the basis of electrophysiology. This description is fully deterministic. Recently the gating mechanisms are also described by stochastic differential equations meaning that the gating is governed by Markov model [31]. This approach has turned out to be very effective for studying the effect of various drugs. The Markov-type model has also been used for describing the genetic effects in ion channels that may lead to cardiac arrhythmia [3].

The propagation of an AP is accompanied by other physical effects in the biomembrane and the axoplasm. The need to understand these effects was already mentioned by Hodgkin [11] and lately in many studies [1, 7]. Indeed, many experiments have shown that the propagation of an AP in a nerve fibre is accompanied by transverse displacements of the biomembrane which means changes in the axon diameter [16, 17, 26, 27, 29]. These local changes called also swelling are small being in the range of 1-2 nm and compared with the diameter of fibres are of several orders smaller. This is confirmed by the more recent experiments by Yang et al. [33]. A typical profile of the transverse displacement is shown in Fig. 4.6.

Fig. 4.6 Profile of the mechanical response (top) in comparison of an AP (bottom) in case of a squid, *Loligo pealei* axon, as measured by Iwasa and Tasaki. Reproduced with permission from [16]; ©Elsevier 1980.

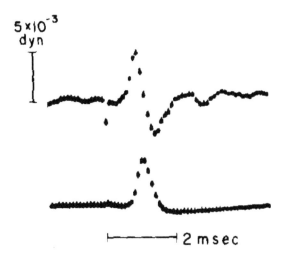

In addition to the deformation of the biomembrane, the pressure wave in the axoplasm has been measured by Terakawa [30]. In a squid axon the amplitude of a pressure wave, measured simultaneously with the AP, was about 1 to 10 mPa. As an example, one of his results is depicted in Fig. 4.7.

The AP and the mechanical processes are accompanied also by the temperature changes. Such changes during the passage of an AP have been measured being of the range about 20-30 μK for a garfish nerve [27] and much less for a bullfrog nerve [28]. The temperature changes are physically related to heat production during the signal propagation [15, 24, etc]. The earlier findings on mechanical and thermal effects are summarised by Watanabe [32] and more recently by Andersen et al. [1]. As an example, the recordings of Tasaki et al. [29] are shown in Fig. 4.8.

The transverse displacement of a cylindrical biomembrane is associated with the longitudinal deformation of the biomembrane. This effect – transverse displacement is proportional to the gradient of the longitudinal displacement – is well understood in mechanics of rods [22]. It means that the bipolar transverse displacement measured by Tasaki [27] corresponds to the unipolar longitudinal displacement and vice versa. The possible deformation of a biomembrane under loading has been studied by measuring the transverse displacement [9, 21] and interpreted then as an accompanying mechanical wave along the biomembrane. This longitudinal wave may have a soliton-type shape [10], i.e., is unipolar.

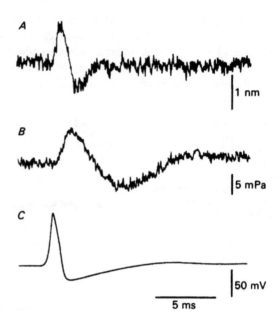

Fig. 4.7 Profiles of the transverse displacement (A) and the pressure wave (B) in comparison to the AP (C) in case of a squid, *Dorytheusis bleekeri* axon, as measured by Terakawa. Reproduced with permission from [30] ; ©John Wiley and Sons 1985.

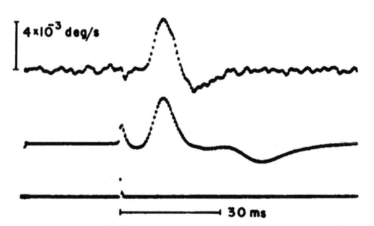

Fig. 4.8 Profile of thermal changes (top) compared to the AP (bottom) in case of a garfish *Lepisosteus osseus* olfactory nerve, as measured by Tasaki et al. [29]. Reproduced with permission from [29] ; ©Elsevier 1989.

It must be noted that the excitable plant cells (*Chara braunii*) behave similarly – the electrical signals are coupled with mechanical effects [8].

To sum up, many physical effects are accompanying the propagation of an AP in nerves. It is a challenge to cast these effects into the language of mathematics based on physical considerations which is done in further Chapters.

References

1. Andersen, S.S.L., Jackson, A.D., Heimburg, T.: Towards a thermodynamic theory of nerve pulse propagation. Prog. Neurobiol. **88**(2), 104–13 (2009). DOI 10.1016/j.pneurobio.2009.03.002
2. Bean, B.P.: The action potential in mammalian central neurons. Nat. Rev. Neurosci. **8**(6), 451–65 (2007). DOI 10.1038/nrn2148
3. Clancy, C.E., Rudy, Y.: Linking a genetic defect to its cellular phenotype in a cardiac arrhythmia Nature **400**(6744), 566–569 (1999). DOI 10.1038/23034
4. Clay, J.R.: Axonal excitability revisited. Prog. Biophys. Mol. Biol. **88**(1), 59–90 (2005). DOI 10.1016/j.pbiomolbio.2003.12.004
5. Courtemanche, M., Ramirez, R.J., Nattel, S.: Ionic mechanisms underlying human atrial action potential properties : insights from a mathematical model. Am. J. Physiol. **275**(1), 301–321 (1998)
6. Debanne, D., Campanac, E., Bialowas, A., Carlier, E., Alcaraz, G.: Axon physiology. Physiol. Rev. **91**(2), 555–602 (2011). DOI 10.1152/physrev.00048.2009.
7. Drukarch, B., Holland, H.A., Velichkov, M., Geurts, J.J., Voorn, P., Glas, G., de Regt, H.W.: Thinking about the nerve impulse: A critical analysis of the electricity-centered conception of nerve excitability. Prog. Neurobiol. **169**, 172–185 (2018). DOI 10.1016/j.pneurobio.2018.06.009
8. Fillafer, C., Mussel, M., Muchowski, J., Schneider, M.F.: Cell surface deformation during an action potential. Biophys. J. **114**(2), 410–418 (2018). DOI 10.1016/j.bpj.2017.11.3776
9. Gonzalez-Perez, A., Mosgaard, L., Budvytyte, R., Villagran-Vargas, E., Jackson, A., Heimburg, T.: Solitary electromechanical pulses in lobster neurons. Biophys. Chem. **216**, 51–59 (2016). DOI 10.1016/j.bpc.2016.06.005
10. Heimburg, T., Jackson, A.D.: On soliton propagation in biomembranes and nerves. Proc. Natl. Acad. Sci. USA **102**(28), 9790–9795 (2005). DOI 10.1073/pnas.0503823102
11. Hodgkin, A.L.: The Conduction of the Nervous Impulse. Liverpool University Press (1964)
12. Hodgkin, A.L., Huxley, A.F.: Action potentials recorded from inside a nerve fibre. Nature **144**(3651), 710–711 (1939). DOI 10.1038/144710a0
13. Hodgkin, A.L., Huxley, A.F.: Resting and action potentials in single nerve fibres. J. Physiol. **104**, 176–195 (1945)
14. Hodgkin, A.L., Huxley, A.F.: A quantitative description of membrane current and its application to conduction and excitation in nerve. J. Physiol. **117**(4), 500–544 (1952). DOI 10.1113/jphysiol.1952.sp004764
15. Howarth, J.V., Keynes, R.D., Ritchie, J.M.: The origin of the initial heat associated with a single impulse in mammalian non-myelinated nerve fibres. J. Physiol. **194**(3), 745–93 (1968). DOI 10.1113/jphysiol.1968.sp008434
16. Iwasa, K., Tasaki, I.: Mechanical changes in squid giant axons associated with production of action potentials. Biochem. Biophys. Res. Commun. **95**(3), 1328–1331 (1980). DOI 10.1016/0006-291X(80)91619-8
17. Iwasa, K., Tasaki, I., Gibbons, R.: Swelling of nerve fibers associated with action potentials. Science **210**(4467), 338–339 (1980). DOI 10.1126/science.7423196
18. de Lichtervelde, A. C. L., de Souza, J. P., Bazant, M. Z.: Heat of nervous conduction: A thermodynamic framework. Phys. Rev. E **101**(2), 022406 (2020). DOI 10.1103/PhysRevE.101.022406

19. Mueller, J.K., Tyler, W.J.: A quantitative overview of biophysical forces impinging on neural function. Phys. Biol. **11**(5), 051001 (2014). DOI 10.1088/1478-3975/11/5/051001
20. Nelson, P.C., Radosavljevic, M., Bromberg, S.: Biological Physics: Energy, Information, Life. W.H. Freeman and Company, New York, NY (2003)
21. Perez-Camacho, M.I., Ruiz-Suarez, J.: Propagation of a thermo-mechanical perturbation on a lipid membrane. Soft Matter **13**, 6555–6561 (2017). DOI 10.1039/C7SM00978J
22. Porubov, A.V.: Amplification of Nonlinear Strain Waves in Solids. World Scientific, Singapore (2003)
23. Ramón, F., Moore, J.W., Joyner, R.W., Westerfield, M.: Squid giant axons. A model for the neuron soma? Biophys. J. **16**(8), 953–963 (1976). DOI 10.1016/S0006-3495(76)85745-1
24. Ritchie, J.M., Keynes, R.D.: The production and absorption of heat associated with electrical activity in nerve and electric organ. Q. Rev. Biophys. **18**(04), 451 (1985). DOI 10.1017/S0033583500005382
25. Sasaki, T.: The axon as a unique computational unit in neurons. Neurosci. Res. **75**(2), 83–88 (2013). DOI 10.1016/j.neures.2012.12.004
26. Tasaki, I., Iwasa, K.: Rapid Pressure Changes and Surface Displacements in the Squid Giant Axon Associated with Production of Action Potentials. Jpn. J. Physiol. **32**(1977), 69–81 (1982). DOI 10.2170/jjphysiol.32.69
27. Tasaki, I.: A macromolecular approach to excitation phenomena: mechanical and thermal changes in nerve during excitation. Physiol. Chem. Phys. Med. NMR **20**(4), 251–268 (1988)
28. Tasaki, I., Byrne, P.M.: Heat production associated with a propagated impulse in Bullfrog myelinated nerve fibers. Jpn. J. Physiol. **42**(5), 805–813 (1992). DOI 10.2170/jjphysiol.42.805
29. Tasaki, I., Kusano, K., Byrne, P.M.: Rapid mechanical and thermal changes in the garfish olfactory nerve associated with a propagated impulse. Biophys. J. **55**(6), 1033–1040 (1989). DOI 10.1016/S0006-3495(89)82902-9
30. Terakawa, S.: Potential-dependent variations of the intracellular pressure in the intracellularly perfused squid giant axon.   J. Physiol. **369**(1), 229–248 (1985).   DOI 10.1113/jphysiol.1985.sp015898
31. Tveito, A., Lines, G.T.: Computing Characterizations of Drugs for Ion Channels and Receptors Using Markov Models. Springer International Publishing, Cham (2016). DOI 10.1007/978-3-319-30030-6
32. Watanabe, A.: Mechanical, thermal, and optical changes of the nerve membrane associated with excitation. Jpn. J. Physiol. **36**, 625–643 (1986). DOI 10.2170/jjphysiol.36.625
33. Yang, Y., Liu, X.W., Wang, H., Yu, H., Guan, Y., Wang, S., Tao, N.: Imaging action potential in single mammalian neurons by tracking the accompanying sub-nanometer mechanical motion. ACS Nano **12**(5), 4186–4193 (2018). DOI 10.1021/acsnano.8b00867.

Chapter 5
Dynamical Effects in Nerves

> *The question of the origin of the hypothesis belongs to a domain*
> *in which no very general rules can be given, experiment,*
> *analogy and constructive intuition play their part here. But once*
> *the correct hypothesis is formulated, the principle of*
> *mathematical induction is often sufficient to provide the proof.*
>
> Richard Courant, 1941

The interplay of electrical and chemical signals in nerves is the basis for neuro-science. As mentioned in the Introduction (Chap. 1), there are many groundbreaking results which have enlarged considerably our knowledge of neural mechanisms. The experiments have revealed that in addition to electrical and chemical signals there are also mechanical and thermal effects accompanying the signalling in nerves. These effects are measurable (see Chap. 4) and the main question is to understand the role of all these physical effects. So the approach to a better understanding of nervous signals is related to grasp the complexity of the problem (see Chap. 2).

The measured characteristics of effects (Chap. 4) give a certain overview on scales but one should take them as estimations. The real situation is very complicated. As said by Debanne et al. [7],

> ... the functional and computational repertoire of the axon is much richer than traditionally thought.

We follow briefly their findings. First, one has to mention that axon morphology is variable in their extension, diameter and arborisation. Further on, several studies have demonstrated [7] that the voltage-gated ion channels neither have uniform properties nor are evenly distributed. The sodium (Na^+), potassium (K^+) and Ca^{2+} channels are of different types regulating specific properties of signals. Conduction velocity is an important characteristic of a signal. In unmyelinated axons which are the topic of our analysis, it depends on several biophysical factors starting from the number of Na^+ channels to the electrical properties of the membrane and temperature. Conduction velocity depends certainly on the geometrical properties of the axon. This was known already to Hodgkin [15]. There are several effects observed during the propagation of an action potential: activity-dependent broadening and reduction of a signal, amplification of a signal along the axon, backward axonal integration, etc. In medical applications, one should also understand the reasons for possible propagation failures and delays imposed by axonal irregularities. Without any doubt, pathologies of axonal function must be understood for explaining reasons for many diseases like epilepsies, multiple sclerosis, etc. Moreover, a better understanding of

J. Engelbrecht et al., *Modelling of Complex Signals in Nerves*,
https://doi.org/10.1007/978-3-030-75039-8_5

nerve signal propagation may help to understand psychiatric disorders like bipolar disorder, schizophrenia, autism, etc. [17, 18].

Heimburg and Jackson [14] have proposed a theory in which the action potential is an electromechanical density pulse (soliton) propagating in a biomembrane. Leaving aside the description of this model, one should stress their detailed studies of the properties of biomembranes. They have established the melting transitions in biomembranes slightly below the body temperature [13]. The density pulse as a transition from physiological conditions of a biomembrane should inhibit electrical pulse. The phase behaviour of biomembranes is strongly influenced by anaesthetics and permits to explain the physical mechanism of the Meyer-Overton rule [23]. This rule says that the action of anaesthetics is proportional to their concentration in lipid membranes. Discovered more than a century ago, the rule is the basis for modern anaesthesiology.

The processes in nerves are based on physical laws. Johnston and Wu [19] have described four fundamental laws of physics which govern the ion movements:
(i) diffusion of particles caused by concentration differences governed by the Fick's law for diffusion;
(ii) the drift of ions governed by potential differences governed by the Ohm's law for drift;
(iii) relationship between the coefficients (diffusion coefficient and drift mobility) of the first two processes governed by the Einstein relation between diffusion and mobility;
(iv) separation of charges in biological systems (space-charge neutrality).

Several models are derived from these basic laws. The Nernst-Planck equation describes the ionic current flow which is driven by the concentration gradient and electric field. The Nernst equation determines the equilibrium potential of ions depending on concentrations inside and outside of the membrane. The Goldman-Hodgkin-Katz model describes the current-voltage relation of the ionic current. These fundamental relationships allow describing the active transport of ions like sodium-potassium (Na^+-K^+) pump, Ca^{2+} pump, etc.

Summarising this brief analysis, two important conclusions can be drawn.

First, all mentioned laws or derived rules described in detail by Johnston and Wu [19] are space dependent. In order to describe the signal propagation in nerves, time dependence must be introduced. That is why wave equations and/or diffusion equation (Chap. 3) must be taken into account.

Second, the processes are extremely complicated and include many variables. It would be an enormous challenge to cast all these phenomena into mathematical language within one general model able to couple all the effects. We follow the recommendation of A. Einstein: "Everything should be made as simple as possible but not simpler". It is easy to say but extremely difficult to realise. The approach used in this book is the following:

(i) Derive time-dependent models (equations) for all the effects which seem to be significant for the whole process based on physical laws;
(ii) Propose coupling mechanisms between the effects;

(iii) Solve the coupled system of equations;
(iv) Validate the results by comparing them with experiments.

In such a simplification process there are certainly many underwater riffs – what should be taken into account and what could be neglected. The mathematical modelling could, however, open new vistas of understanding like explained in Chap. 2 based on the Report from the National Research Council [22] describing bio-mathematics. In some sense, such an approach could be compared with the idea of E. Lorenz [20] who has derived his famous system of three ordinary differential equations from the general Navier-Stokes equations describing the processes in the atmosphere.

Further, we deal with an **ensemble of waves** in the axon. This ensemble has the following components (notations correspond to the dimensionless case):
(i) **the action potential AP** which has an amplitude Z and the ion currents. In case of the Hodgkin-Huxley (HH) model, these ion currents are J_K, J_{Na} and J_L, in case of the FitzHugh-Nagumo (FHN) model just one ion current J (here the ion currents for the HH model denote potassium, sodium and leakage currents, respectively).
(ii) **the longitudinal wave LW** in the biomembrane with an amplitude U;
(iii) **the transverse displacement TW** in the biomembrane with an amplitude W;
(iv) **the pressure wave PW** in the axoplasm with an amplitude P;
(v) **the temperature change** Θ.

The AP depends on the properties of the axoplasm and ion currents, the PW depends on the properties of the axoplasm. The LW and the TW depend on the elastic properties of the biomembrane, while the Θ depends on the thermodynamical parameters of the whole system. All these quantities should be coupled into a whole which reflects the complexity of the process of signal propagation in an axon. In Fig. 5.1 a scheme of such an ensemble is presented. A sketch with profiles of waves in the ensemble is shown in Fig. 5.2.

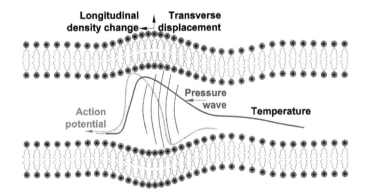

Fig. 5.1 A scheme of the components of a signal in the axon (scales given in the text).

Fig. 5.2 A scheme of the
ensemble of waves. See text
for notations. The scales are
arbitrary. The amplitudes are:
AP ~100 mV, PW 1-10 mPa,
TW 1-2 nm, Θ 10-50 μK.

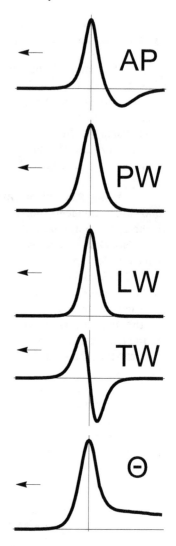

Although molecular mechanisms (ion currents) influence the process, the values
of the main variables and axon parameters are far from the molecular range. In terms
of continuum theories, this means the micro- or mesoscale [11] which justifies the
usage of models derived from basic conservation laws of continua. Starting from the
basics (see Chap. 3), the wave equations are the cornerstones of all the processes
which propagate in media with certain finite velocities. The diffusion equations
describe the processes related to heat production. Given the complicated structure
of an axon, these model equations should be modified.

Based on the classical understandings of axon physiology [6, 7], the following
basic assumptions are made:

(i) electrical signals are the carriers of information [7] and trigger all the other processes;

(ii) the axoplasm in a fibre can be modelled as a fluid where a pressure wave is generated due to the electrical signal; here, for example, the actin filaments in the axoplasm may influence the opening of channels in the surrounding biomembrane but do not influence the generation of a pressure wave in the fluid [1];

(iii) the biomembrane is able to deform (stretch, bending) under the mechanical impact [14];

(iv) the ion channels in biomembranes can be opened and closed under the influence of electrical signals as well as of the mechanical input; it means that tension of a membrane leads to the increase of transmembranal ion flow and the intracellular actin filaments may influence the motions at the membrane [3, 21];

(v) there is strong experimental evidence on electrical or chemical transmittance of signals from one neuron to another [4, 16] although the role of the mechanical transmission is also discussed [2].

In electrophysiology, such an approach is sometimes called the Hodgkin-Huxley paradigm. A different approach to signal generation is proposed by Heimburg and Jackson [14] where the mechanical wave in the biomembrane (LW) has the leading role (see more in Chap. 8).

The trivial idea for constructing a mathematical model is to collect all the single equations describing the processes in a nerve fibre but these should be coupled into a joint system. The question of how the coupling forces between the equations (i.e., signal components) are modelled is crucial. It is not only a mathematical construct but the forces reflect the electro-mechanical (or mechano-electrical) transduction mechanisms.

We introduce the **main hypothesis** [9]:
all mechanical waves in axoplasm and surrounding biomembrane together with the heat production are generated due to changes in electrical signals (AP or ion currents) that dictate the functional shape of coupling forces.

The **second hypothesis** is [10]:
the formalism of internal variables can be used for describing the exo- and endothermic processes of heat production.

The **additional hypothesis** is: The changes in the pressure wave may also influence the waves in a biomembrane.

It must be noted that already in the 19th century the German physiologist Emil Du Bois-Reymond has noticed that [12]

the variation of current density, and not the absolute value of the current density at any given time, acts as a stimulus to a muscle or motor nerve.

This statement is called the Du Bois-Reymond law.

Two essential remarks are following: (i) the **changes of variables** mean mathematically either **space or time derivatives**; (ii) the pulse-type profiles of electrical signals mean that the derivatives have a **bi-polar shape** which is energetically balanced. Based on these remarks, the functional shapes of coupling forces are proposed in the form of first-order polynomials of gradients or time derivatives of variables [8, 9].

Due to reciprocity, the mechanical effects could also influence the electrical properties of the membrane and consequently the behaviour of the AP [5]. In mathematical terms, it means that the model should have the variable-dependent parameters. In the present model, the feedback from mechanical effects to the AP is proposed through introducing the dependence on LW into the ion current. It is possible to include some scale factors into the coupling forces but such a proposal needs more studies in order to prove a need for such an assumption. As the dynamical changes in the axon diameter (the amplitude of the TW) are many orders less than the diameter itself [26], the electrical properties of the fibre are taken constant during all the process.

In addition, it is assumed that the TW is derived from the LW like in mechanics of rods [24] – the transverse displacement is a space derivative of the longitudinal displacement. Concerning the possible temperature changes, there are several possibilities to build up a mathematical model [25].

It must be stressed that the assumptions described above are logically consistent within the present framework of axon electrophysiology [6, 7] and experimental studies on accompanying effects. What is important – all the components of a wave ensemble including the AP are coupled. The model serves the contemporary trends in computational biology and could help experimentalists to check the possible transduction mechanism [22].

References

1. Andersen, S.S., Jackson, A.D., Heimburg, T.: Towards a thermodynamic theory of nerve pulse propagation. Prog. Neurobiol. pp. 104–113 (2009). DOI 10.1016/j.pneurobio.2009.03.002
2. Barz, H., Barz, U.: Pressure waves in neurons and their relationship to tangled neurons and plaques. Med. Hypotheses **82**(5), 563–566 (2014). DOI 10.1016/j.mehy.2014.02.012
3. Barz, H., Schreiber, A., Barz, U.: Impulses and pressure waves cause excitement and conduction in the nervous system. Med. Hypotheses **81**(5), 768–772 (2013). DOI 10.1016/j.mehy.2013.07.049
4. Bennett, M.V.: Electrical synapses, a personal perspective (or history). Brain Res. Rev. **32**(1), 16–28 (2000). DOI 10.1016/S0165-0173(99)00065-X
5. Chen, H., Garcia-Gonzalez, D., Jérusalem, A.: Computational model of the mechanoelectro-physiological coupling in axons with application to neuromodulation. Phys. Rev. E **99**(3), 032406 (2019). DOI 10.1103/PhysRevE.99.032406
6. Clay, J.R.: Axonal excitability revisited. Prog. Biophys. Mol. Biol. **88**(1), 59–90 (2005). DOI 10.1016/j.pbiomolbio.2003.12.004
7. Debanne, D., Campanac, E., Bialowas, A., Carlier, E., Alcaraz, G.: Axon physiology. Physiol. Rev. **91**(2), 555–602 (2011). DOI 10.1152/physrev.00048.2009.
8. Engelbrecht, J., Peets, T., Tamm, K., Laasmaa, M., Vendelin, M.: On the complexity of signal propagation in nerve fibres. Proc. Estonian Acad. Sci. **67**(1), 28–38 (2018). DOI 10.3176/proc.2017.4.28
9. Engelbrecht, J., Tamm, K., Peets, T.: Modeling of complex signals in nerve fibers. Med. Hypotheses **120**, 90–95 (2018). DOI 10.1016/j.mehy.2018.08.021
10. Engelbrecht, J., Tamm, K., Peets, T.: Internal variables used for describing the signal propagation in axons. Contin. Mech. Thermodyn. **32**(6), 1619–1627 (2020). DOI 10.1007/s00161-020-00868-2

11. Gates, T.S., Odegard, G.M., Frankland, S.J., Clancy, T.C.: Computational materials: Multi-scale modeling and simulation of nanostructured materials. Compos. Sci. Technol. **65**(15-16), 2416–2434 (2005). DOI 10.1016/j.compscitech.2005.06.009
12. Hall, C.W.: Laws and Models: Science, Engineering, and Technology. CRC Press, Boca Raton (1999)
13. Heimburg, T., Jackson, A.: On the action potential as a propagating density pulse and the role of anesthetics. Biophys. Rev. Lett. **2**, 57–78 (2007)
14. Heimburg, T., Jackson, A.D.: On soliton propagation in biomembranes and nerves. Proc. Natl. Acad. Sci. USA **102**(28), 9790–9795 (2005). DOI 10.1073/pnas.0503823102
15. Hodgkin, A.L.: A note on conduction velocity. J. Physiol. **125**(1), 221–224 (1954). DOI 10.1113/jphysiol.1954.sp005152
16. Hormuzdi, S.G., Filippov, M.A., Mitropoulou, G., Monyer, H., Bruzzone, R.: Electrical synapses: a dynamic signaling system that shapes the activity of neuronal networks. Biochim. Biophys. Acta - Biomembr. **1662**(1-2), 113–137 (2004). DOI 10.1016/j.bbamem.2003.10.023.
17. Huang, C. Y.-M., Rasband, M. N.: Axon initial segments: structure, function, and disease. Ann. N. Y. Acad. Sci. **1420**(1), 46–61 (2018). DOI 10.1111/nyas.13718
18. Imbrici, P., Camerino, D. C., Tricarico, D.: Major channels involved in neuropsychiatric disorders and therapeutic perspectives. Front. Genet. **4**, 1–19 (2013). DOI 10.3389/fgene.2013.00076
19. Johnston, D., Wu, S.M.S.: Foundations of Cellular Neurophysiology. The MIT Press, Cambridge, Mass. (1995)
20. Lorenz, E.N.: Deterministic nonperiodic flow. J. Atmos. Sci. **20**(2), 130–141 (1963). DOI 10.1175/1520-0469(1963)020<0130:DNF>2.0.CO;2
21. Mueller, J.K., Tyler, W.J.: A quantitative overview of biophysical forces impinging on neural function. Phys. Biol. **11**(5), 051001 (2014). DOI 10.1088/1478-3975/11/5/051001
22. National Research Council: Catalyzing Inquiry at the Interface of Computing and Biology. The National Academies Press, Washington (2005). DOI 10.17226/11480
23. Overton, C.E.: Studies of Narcosis. Chapman & Hall (1991)
24. Porubov, A.V.: Amplification of Nonlinear Strain Waves in Solids. World Scientific, Singapore (2003)
25. Tamm, K., Engelbrecht, J., Peets, T.: Temperature changes accompanying signal propagation in axons. J. Non-Equilibrium Thermodyn. **44**(3), 277–284 (2019). DOI 10.1515/jnet-2019-0012
26. Tasaki, I.: A macromolecular approach to excitation phenomena: mechanical and thermal changes in nerve during excitation. Physiol. Chem. Phys. Med. NMR **20**(4), 251–268 (1988)

Part III
Modelling of Dynamical Physiological Processes

Chapter 6
Mathematics of Single Effects

> *Mathematics is not about numbers, equations, computations, or algorithms: it is about understanding.*
>
> William Paul Thurston, 1994

The ensemble of waves in an axon is described by governing equations of single effects. Before building up a coupled system, the properties of those governing equations and their possible solutions must be well understood. Further, we collect the results of the analysis where the main attention is paid to the AP and the LW.

6.1 The Action Potential

The physiological background of the AP is described in Chap. 4. In this Section, we will turn our attention to mathematical models describing the AP.

6.1.1 The Classical Hodgkin-Huxley Model

The celebrated Hodgkin-Huxley (HH) model is a cornerstone in contemporary understanding of axon physiology. Proposed in the mid-20th century by A.L. Hodgkin and A.F. Huxley [28, 29, 51], it describes explicitly the role of ion currents in forming an asymmetric AP in an unmyelinated nerve fibre. The experiments with the squid axons permitted to determine a model with experimentally verified coefficients. Here we represent the main description of the HH model [41, 52].

The starting point for the derivation of the HH model stems from the hyperbolic telegraph equations. Following Lieberstein [31] these equations are the following (in original notations)

$$\pi a^2 C_a \frac{\partial v}{\partial t} + \frac{\partial i_a}{\partial x} + 2\pi a I = 0,$$
$$\frac{L}{\pi a^2} \frac{\partial i_a}{\partial t} + \frac{\partial v}{\partial x} + \frac{R}{\pi a^2} i_a = 0,$$

(6.1)

© The Author(s), under exclusive license to Springer Nature Switzerland AG 2021
J. Engelbrecht et al., *Modelling of Complex Signals in Nerves*,
https://doi.org/10.1007/978-3-030-75039-8_6

where v is the potential difference across the biomembrane, i_a is the axon current per unit length and I is ion current density; the constants are: C_a is the axon self-capacitance per unit area per unit length, L is the axon specific self-inductance and R is the axon specific resistance. As usual, x and t are space coordinate and time, respectively and a is the radius of an axon.

It is possible to rewrite system (6.1) in the form of one second order equation

$$\frac{\partial^2 v}{\partial x^2} - LC_a \frac{\partial^2 v}{\partial t^2} = RC_a \frac{\partial v}{\partial t} + \frac{2}{a} RI + \frac{2}{a} L \frac{\partial I}{\partial t}. \tag{6.2}$$

Since in electrophysiology inductance L is usually assumed to be negligible, then the Eq. (6.2) can be rewritten as

$$\frac{\partial^2 v}{\partial x^2} = RC_a \frac{\partial v}{\partial t} + \frac{2}{a} RI. \tag{6.3}$$

It is important to note that Eq. (6.2) is a hyperbolic and Eq. (6.3) is a parabolic partial differential equation (PDE). The main difference between the two is that in case of a hyperbolic PDE the disturbances travel at finite and in case of a parabolic PDE at infinite speeds and consequently equations describing travelling waves are usually hyperbolic PDEs (see Chap. 3). In case of Eq. (6.3), the finite speed of a travelling wave is ensured by the ion current $j = 4\pi a I$. Such equations are known as reaction-diffusion type equations and consequently we refer to these type of equations as wave-like equations [4, 16].

Hodgkin and Huxley [28] realised from the experiments that the membrane current I can be divided into four components – the capacitive current I_C, ion currents of potassium I_K and sodium I_{Na} and a leakage current I_L which consists of chloride and other ions. Denoting the maximum conductances of these ion components as g_{Na}, g_K and g_L then the membrane current expression is deduced

$$I = g_K n^4 (v - v_K) + g_{Na} m^3 h (v - v_{Na}) + g_L (v - v_L) + C_m \frac{\partial v}{\partial t}, \tag{6.4}$$

where C_m is the membrane capacitance per unit area and v_{Na}, v_K, v_L denote equilibrium potentials of the corresponding ions. The phenomenological (hidden) variables n, m, h govern the 'turning on' and 'turning off' individual membrane conductances – variable n governs the potassium conductance ('turning on') and the variables m, h govern the sodium conductance ('turning on' and 'turning off', respectively). The values of these variables are calculated from the following kinetic equations:

$$\frac{dn}{dt} = \alpha_n (1 - n) - \beta_n n,$$
$$\frac{dm}{dt} = \alpha_m (1 - m) - \beta_m m, \tag{6.5}$$
$$\frac{dh}{dt} = \alpha_h (1 - h) - \beta_h h.$$

The expression for the parameters in kinetic equations are determined experimentally[29]:

$$\alpha_n = \frac{0.01(10 - V)}{\exp[(10 - V)/10] - 1},$$

$$\beta_n = 0.125 \exp(-V/80),$$

$$\alpha_m = \frac{0.1(25 - V)}{\exp[(25 - V)/10] - 1},$$

$$\beta_m = 4 \exp(-V/18),$$

$$\alpha_h = 0.07 \exp(-V/20),$$

$$\beta_h = \frac{1}{\exp[(30 - V)/10] + 1}.$$

(6.6)

These coefficients are in units of $(ms)^{-1}$ while potential V is in mV.

The celebrated HH model has been tested by many experiments. Several modifications have been proposed to modify the ion current j (see, for example, [8]). An excellent overview on the HH model is presented by Sterratt et al. [53]. However, the HH model describes the AP like an electrical signal supported by ion currents and does not take into account other accompanying effects [3].

The HH model is able to describe several effects of an AP like listed in Chap. 4. The calculations demonstrate that the AP is asymmetric with the refractory period (c.f. Fig. 4.4) and the corresponding sodium and potassium ion currents have opposite signs. This is shown in Fig. 6.1.

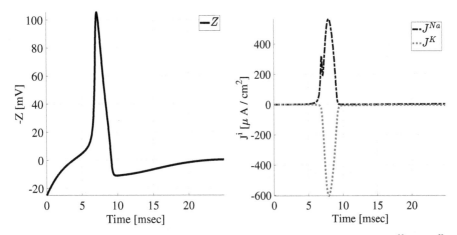

Fig. 6.1 Left panel: the AP (amplitude Z) and right panel: individual ion currents J^{Na} and J^K from the classical HH model [29]. Note that the relevant figure (Fig. 18 in [29]) has been found at $18.5°C$ while the lines in the present figure have been found using the reference values given for the model (Table 3 in [29]) which are at $6.3°C$.

If other ion currents dominate then the HH model must be modified. Morris and Lecar [37] have derived a model for the barnacle (*balanus nubilus*) giant muscle fibre. In this case the K and Ca^{2+} currents are taken into account. The Morris-Lecar (ML) model uses the membrane current expression in the form (c.f. expression (6.4)):

$$I = g_L(V_L) + g_{Ca}M(V - V_{Ca}) + g_K N(V - V_K) + C_m \frac{\partial V}{\partial t}. \tag{6.7}$$

Here the original notations are used where V is the membrane potential; V_L, V_{Ca}, V_K are the equilibrium potentials corresponding to leak, Ca^{2+} and K conductances, respectively; g_L, g_{Ca}, g_K – are the corresponding maximum conductances; C is the membrane capacitance. Instead of the phenomenological variables n, m, h on the HH model (6.4) here two variables M and N are used corresponding to m and n in the HH model. Physically M denotes the fraction of open Ca^{2+} channels and N - those of open K channels. They are governed by (c.f. expressions (6.5)):

$$\begin{aligned}
\frac{dM}{dt} &= \lambda_M(V)[M_\infty(V - M)], \\
\frac{dN}{dt} &= \lambda_N(V)[N_\infty(V - N)],
\end{aligned} \tag{6.8}$$

where λ_M, λ_N are rate constants, M_∞ and N_∞ – fractions of open Ca^{2+} and K channels at steady state. These quantities are governed by expressions:

$$\begin{aligned}
M_\infty(V) &= \frac{1}{2}\{1 + \tanh[(V - V_1)/V_2]\}, \\
\lambda_\infty(V) &= \lambda_M \cosh[(V - V_1)/2V_2], \\
N_\infty(V) &= \frac{1}{2}\{1 + \tanh[(V - V_3)/V_4]\}, \\
\lambda_\infty(V) &= \lambda_N \cosh[(V - V_3)/2V_4],
\end{aligned} \tag{6.9}$$

where V_1, V_2, V_3, V_4 are related to potential for certain values of M_∞ and N_∞ and to the slope of potential. The ML model leads to voltage oscillations. Note however that the structure of the ML model is similar to the HH model but the differences are in describing the phenomenological variables.

6.1.2 The FitzHugh-Nagumo Model

The existence of many physical parameters in the HH model (see above) may cause problems in modelling. That is why the simplified models are sought that still could describe the main effects. For example, FitzHugh [23] used the ideas of Bonhoeffer [5] and van der Pol [46] for deriving a model of excitable-oscillatory system. Actually, FitzHugh named his model after Bonhoffer and van der Pol but starting from the paper by Nagumo et al. [38], the model is called after FitzHugh and Nagumo. Here and below the notations from original papers are used. The starting point is the van der Pol equation

$$\ddot{x} + c(x^2 - 1)\dot{x} + x = 0, \tag{6.10}$$

where x is an oscillating quantity (amplitude) and c is a positive constant. The dots denote differentiation with respect to time t. By using Liénard's transformation

$$y = \dot{x}/c + x^3/3 - x, \tag{6.11}$$

the system of the first order equations is obtained:

$$\begin{aligned} \dot{x} &= c(y + x - x^3/3), \\ \dot{y} &= -x/c. \end{aligned} \tag{6.12}$$

The Bonhoeffer-van der Pol (BVP) model enlarges system (6.12) into

$$\begin{aligned} \dot{x} &= c(y + x - x^3/3 + z), \\ \dot{y} &= -(x - a + by)/c, \end{aligned} \tag{6.13}$$

where

$$1 - 2b/3 < a < 1, \ \ 0 < b < 1, \ \ b < c^2. \tag{6.14}$$

Here a and b are constants, z is the stimulus intensity representing the membrane current.

It is noted that x is related to membrane potential and excitability and y – to accommodation and refractoriness. The phase-plane analysis is used for demonstrating the threshold and all-or-none phenomena together with the refractory part of the solution. Numerical solution using $a = 0.7, b = 0.8, c = 3$ and various values of z are obtained ([23], Fig. 2).

In terms of variables, the variables x, y, z in the BVP model correspond to variables v and m, h and n, I in the HH model, respectively. However, the HH model describes the propagating wave (Eq. (6.3) is a PDE) while the BVP model describes the standing (stationary) profile (Eqs. (6.13) are ODE's).

In original notations the FitzHugh-Nagumo (FHN) model is written like the BVP model (6.13) in the form of a system [38]:

$$\begin{aligned} h\frac{\partial^2 u}{\partial s^2} &= \frac{1}{c}\frac{\partial u}{\partial t} - w - \left(u - \frac{u^3}{3}\right), \\ c\frac{\partial w}{\partial t} + bw &= a - u, \end{aligned} \tag{6.15}$$

where u is the voltage and w – the recovery current. The constants h, c, b, a are positive satisfying

$$1 > b > 0, \ \ c^2 > b, \ \ 1 > a > 1 - \frac{2}{3}b. \tag{6.16}$$

As noted by Nagumo et al. [38], system (6.15) is the distributed BVP model (6.13) which is an ODE. The possible governing equation can be derived from (6.15)

$$ch\frac{\partial^3 u}{\partial t \partial s^2} = \frac{\partial^2 u}{\partial t^2} - c\left(1 - u^2\right)\frac{\partial u}{\partial t} + u - a, \tag{6.17}$$

where for simplicity $b = 0$.

By setting $x = s/\sqrt{ch}$, $z = 2a(a - u)/(a^2 - 1)$, $\mu = c(a^2 - 1)$, $\varepsilon = (a^2 - 1)/(4a^2)$, the FHN equation is deduced [38]:

$$\frac{\partial^3 z}{\partial t \partial x^2} = \frac{\partial^2 z}{\partial t^2} + \mu\left(1 - z + \varepsilon z^2\right)\frac{\partial z}{\partial t} + z, \tag{6.18}$$

where $\mu > 0$ and $3/16 > \varepsilon > 0$.

A modification of Eqs. (6.15) changes the polynomial $u - 1/3u^3$ to its full form [6, 39]. Then Eqs. (6.15) can be represented in the form of two coupled equations [17] using dimensionless variables z and j:

$$\begin{aligned} \frac{\partial z}{\partial t} &= z(z - a)(1 - z) - j + D\frac{\partial^2 z}{\partial x^2}, \\ \frac{\partial j}{\partial t} &= \varepsilon(-j + bz), \end{aligned} \tag{6.19}$$

where D is a coefficient, ε is the time-scale difference, activation parameters satisfy conditions $0 < a < 1$ and $b > 0$.

The FHN model described above is simpler to analyse as it involves only one ion current instead of individual ion currents tied to the kinetic equations as it is done in case of the HH model. Nowadays the FHN model is one of the cornerstones in the analysis of the nerve pulse and/or more generally, the dynamics of excitable media [14].

As an example a numerical solution of system (6.19) is shown in Fig. 6.2. It can be seen that the FHN model gives a correct shape of the AP (solid black). The ion current is plotted in dashed red. The profiles depicted here and further propagate from right to left.

The important feature of nerve pulses is the annihilation of two colliding pulses. This can also be demonstrated with the FHN model. Figure 6.3 shows a simulation of this process. Two pulses are initiated at different spatial points and the pulses completely annihilate upon collision.

The FHN model is also able to describe the train of pulses. For example, if initial pulses are applied periodically in time at $T_i = i\Delta T$ with $i = 0, 1, 2, \ldots$, a train of periodic pulses is formed. The time between pulses ΔT is equal to the refraction time and it is determined from numerical experiment. If an initial pulse is applied between time interval $T_i < T < T_{i+1}$, then a propagating pulse is not formed. This is in good agreement with experiments. Such an effect is demonstrated in Fig. 6.4. The train of two pulses is shown in Fig. 6.4 bottom panel at the location indicated by red dotted line in time slice plot (top panel).

Fig. 6.2 Solution of FHN equation (6.19): dimensionless membrane voltage Z is plotted in solid black and dimensionless ion current J in dashed red. Here the following parameters are used: $D = 1$, $a = b = -0.2$ and $\varepsilon = 0.018$.

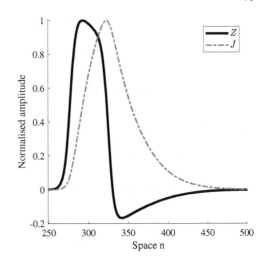

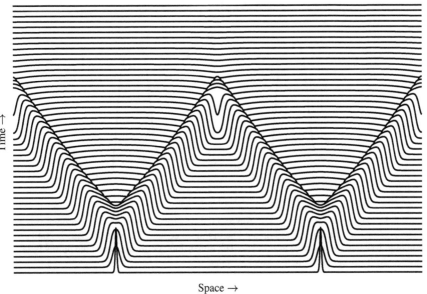

Fig. 6.3 Time slice plot showing annihilation of colliding pulses. Initial pulses $Z(X, 0)$ have amplitude 1.2, width parameter $B = 1$. The size of space is $n = 2048$ and time separation between each horizontal line is 10 units.

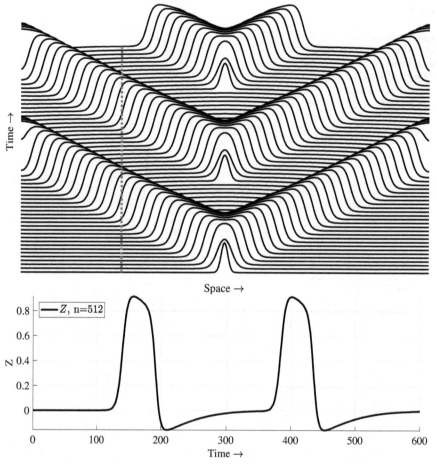

Fig. 6.4 Formation of a train of pulses. Top panel: time slice plot, bottom panel: train of two pulses at the location indicated by the red dotted line in time slice plot. The size of space is 2048 and the train of pulses in time is shown at $n = 512$. The dimensionless time between the pulses is 240 and between the horizontal lines in top panel 10 units.

6.1.3 The Evolution Equation

From wave mechanics, it is known that the classical wave equation describes two waves – one propagating to the right, another – to the left. Under certain conditions, these waves can be separated and in this case, the result is an evolution equation which describes one wave – either propagating to the right or to the left. In mathematical terms, the leading derivative is then of the first order. The details of such derivation by the reductive perturbation method are described in several monographs [13, 56].

Models governing the propagation of the AP are as a rule derived from the telegraph equation where the inductance is neglected. The result is a parabolic equation which due to the existence of the ion current j leads to the propagating wave. The full telegraph equation involves also the inductance L and consequently, leads to the finite velocity $c_0^2 = a/(2LC)$. This velocity is not the velocity of propagation as it is strongly influenced by the ion current. Hyperbolic extension of the HH equation has been derived by Lieberstein [31]. On the other side, the existence of the finite velocity in the initial model permits to use the approaches known for deriving the evolution equations [13, 14, 56].

The derivation of an evolution equation means the splitting the wave process into single waves: in the case of the one-dimensional process instead of two waves governed by the conventional wave equation, the evolution equation governs the propagation of one wave only, either to the right or to the left. For example, the well-known Korteweg-de Vries and Burgers equations are widely used one-wave models.

The evolution equation for the nerve pulse has been derived by Engelbrecht [12]. The starting point of the derivation is the full telegraph equation (6.2) and a simplified description of the ion current like in case of the FHN model. The evolution equation in a moving frame $\xi = c_0 t - x$ is obtained in the following form [12, 14]:

$$\frac{\partial^2 z}{\partial x \partial \xi} + f(z)\frac{\partial z}{\partial \xi} + g(z) = 0, \tag{6.20}$$

where

$$f(z) = \mu(b_0 + b_1 z + b_2 z^2), \quad g(z) = b_{00} z. \tag{6.21}$$

Here $\mu, b_0, b_1, b_2, b_{00}$ are constants.

Like the HH and FHN models, Eq. (6.20) is able to reflect the main properties of the action potential – the existence of a threshold for propagating the signal, the existence of a steady (stationary) pulse and the existence of a refraction length. The moving frame includes the velocity c_0 determined from the telegraph equation but the final velocity of a pulse is dictated by the ion current. The steady pulse is described by an ODE in terms of $\eta = x + \Lambda \xi$, where $\Lambda > 0$ reflects the inclination from the velocity c_0. This equation reads:

$$z'' + f(z)z' + \Lambda^{-1} g(z) = 0, \tag{6.22}$$

where $z' = dz/d\eta$.

Equation (6.22) is a Liénard-type equation [49] which governs the stationary pulse. Assuming $b_3 > 0$ and noting that its behaviour depends on roots z_1, z_2, of $f(z) = 0$. If $z_1 = -z_2 \neq 0$, then Eq. (6.22) is the van der Pol equation; if $z_1 < 0$, $z_2 > 0$, then Eq. (6.22) describes the process in a lamp generator with a 'soft regime' [2]. Using the Bendixson's theorem, it is possible to show that the van der Pol equation has a limit cycle as demonstrated already by the author himself long time ago but under the conditions of a lamp generator there is no limit cycle [22].

Equation (6.22) can also be presented as a system

$$dz/d\eta = y,$$

$$dy/d\eta = -b_4 z - (b_0 + b_1 z + b_2 z^2)\mu, \tag{6.23}$$

where $b_4 = b_{00}/\Lambda$.

Fig. 6.5 Phase plot (top panel) and a solution (bottom panel) of system (6.23). Here $\mu = 2$, $b_{00} = 2.6$, $b_0 = 1$, $b_1 = -1$ and $b_2 = 0.1$.

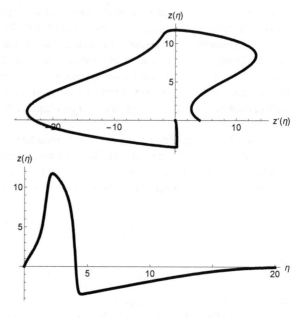

It is easy to see that if $b_3 \neq 0$ then the origin $(0,0)$ is the only singular point of the system. If $b_0^2 - 4b_4 > 0$ then this singular point is a node [57]. The phase trajectories are determined by the equation

$$dy/dz = -(b_0 + b_1 z + b_2 z^2) - b_4 z/y \tag{6.24}$$

and the zero-isoclines – by the equation

$$h(z) = -b_4 z/(b_0 + b_1 z + b_2 z^2). \tag{6.25}$$

Details about the solutions of Eq. (6.22) can be found in [14]. As an example a phase plot and solution of system (6.23) is shown in Fig. 6.5, where it can be seen that the Liénard-type equation is able to grasp the main shape of the AP.

6.2 The Longitudinal Wave in a Biomembrane

As explained in Chap. 4, the propagation of the AP is accompanied also by dynamical deformation of the biomembrane. A mathematical model for the longitudinal deformation is proposed by Heimburg and Jackson [26] and later improved by Engelbrecht et al. [19]. As far as this model is one of the basic elements of a nerve fibre, the properties of it should be explicitly analysed. We present here the results based on the analysis of Engelbrecht et al. [21] and Peets et al. [41, 44] (see Acknowledgments).

6.2.1 Model Derivation

The Heimburg-Jackson (HJ) model for longitudinal wave in biomembrane is based on the classical wave equation and on two assumptions. The classical wave equation in terms on density change $\Delta\rho_A = u$ is

$$u_{tt} = (c_e^2 u_x)_x. \tag{6.26}$$

The first assumption is that the compressibility of a biomembrane has effect on the velocity c_e in a biomembrane. To a first approximation, this impact is modelled by a quadratic function as

$$c_e^2 = c_0^2 + pu + qu^2, \tag{6.27}$$

where c_0 is a velocity in unperturbed state and p, q are coefficients determined from experiments.

The second assumption is that the propagation of sound in a biomembrane is dispersive (that is, phase velocity is not constant but is a function of the wave number). This is modelled by adding a *ad hoc* term $-hu_{xxxx}$ to the governing equation responsible for dispersion [26]. The governing equation is then

$$u_{tt} = \left[(c_0^2 + pu + qu^2)u_x\right]_x - hu_{xxxx}, \tag{6.28}$$

where h is a constant.

Equation (6.28) is a Boussinesq-type equation [7] grasping the following effects: (i) bi-directionality of waves, (ii) nonlinearity (of any order) and (iii) dispersion (of any order modelled by space and time derivatives of the fourth order at least). In case of the HJ model (6.28), there is only one dispersion term hu_{xxxx} which reflects the elastic properties of a biomembrane. However, it is known from theory and experiments of microstructured materials (see, for example, [34, 35, 47, etc.]) that proper modelling should also consider the inertial effects.

Following the ideas from the solid mechanics and supported by the Lagrangian formalism, this shortcoming of the HJ model (6.28) was addressed by Engelbrecht et al. [19] by means of inserting an additional dispersion term $h_2 u_{xxtt}$ (which is related to the inertial properties of the biomembrane) into this equation. The improved HJ (iHJ) equation reads:

$$u_{tt} = \left[(c_0^2 + pu + qu^2)u_x\right]_x - h_1 u_{xxxx} + h_2 u_{xxtt}, \tag{6.29}$$

where $h_1 = h$ and h_2 are dispersion coefficients.

The importance of the additional dispersion term $h_2 u_{xxtt}$ can be explained by dispersion analysis. It has been demonstrated [19] that in case of only one dispersion term $h_1 u_{xxxx}$ (Eq. (6.28)), the phase velocity is expressed as $c_{ph}^2 = c_0^2 + h_1 k^2$ and it tends to infinity as the wave number k is increased. If the second fourth order mixed dispersion term $h_2 u_{xxtt}$ is added then the propagation velocity is bounded as it can be seen in Fig. 6.6. The bounding velocity c_1 for high frequency harmonics is defined by the ratio of the dispersion coefficients ($c_1^2 = h_1/h_2$) while coefficient h_2 is related to the rate of change of the velocity from low frequency to the high frequency domain. Higher valued coefficient h_2 means that the transition from the low frequency speeds to the higher frequency speed is more rapid (see Fig. 6.6). We also note that $c_1/c_0 < 1$ means normal dispersion, i.e., the higher frequency harmonics travel slower than the lower frequency harmonics and $c_1/c_0 > 1$ means anomalous dispersion.

From the viewpoint of solid mechanics it must be stressed that the importance of the fourth order mixed derivative is not surprising as it is well known that the presence of only spatial derivatives in the governing equation can lead to instabilities [32]. Moreover, the mixed fourth order derivative is related to the inertia of the microstructure. It is shown by Maurin and Spadoni [34] that both dispersive terms arise naturally as a result of proper modelling and this has also been demonstrated experimentally [35].

Fig. 6.6 Phase speed curves for Eq. (6.29) in case of normal (top panel) and anomalous dispersion (bottom panel). Parameters used: $c_1/c_0 = 0.9$ (top panel) and $c_1/c_0 = 1.1$ (bottom panel); $h_2/c_0^2 = 1$ (dashed lines) and $h_2/c_0^2 = 0.15$ (solid lines) for both cases.

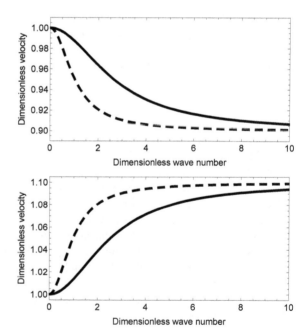

6.2.2 Steady Solutions

For the convenience of further analysis Eq. (6.29) is written in a dimensionless form [43, 44]:

$$U_{TT} = \left[(1 + PU + QU^2)U_X\right]_X - H_1 U_{XXXX} + H_2 U_{XXTT}, \qquad (6.30)$$

with $U = u/\rho_A$, $X = x/l$, $T = c_0 t/l$, $P = p\rho_A/c_0^2$, $Q = q\rho_A^2/c_0^2$, $H_1 = h_1/(c_0^2 l^2)$ and $H_2 = h_2/l^2$. Here l is a certain length, for example, the diameter of the axon.

Solution that propagates at constant dimensionless velocity c while preserving its shape can be written as $V = V(\xi)$ where V is some function and $\xi = X - cT$ is a moving frame [1, 9, 11]. Substituting this ansatz into Eq. (6.30) and integrating twice, a second order ODE [21, 43]

$$(H_1 - H_2 c^2)V'' = (1 - c^2)V + \frac{1}{2}PV^2 + \frac{1}{3}QV^3 + A\xi + B \qquad (6.31)$$

is obtained, where A, B are constants of integration and $(\ )' = d/d\xi$. This equation can be integrated with standard ODE solvers for numerical solutions. Further we will focus on solitary wave solutions and require that $V, V', V'' \to 0$ as $X \to \pm\infty$ and therefore $A, B = 0$ [1, 9, 11].

For deriving analytic solutions, Eq. (6.31) is multiplied by V' and integrated once to get

$$(H_1 - H_2 c^2)(V')^2 = (1 - c^2)V^2 + \frac{1}{3}PV^3 + \frac{1}{6}QV^4. \qquad (6.32)$$

Using an useful analogy form mechanics, Eq. (6.32) can be interpreted as conservation of energy with the lhs acting as a kinetic term and the rhs as a 'pseudopotential' $\Phi_{eff}(V)$ [1, 9, 11].

Next, Eq. (6.32) is rewritten as

$$(V')^2 = V^2 \left[(V - a_+)(V - a_-)\right] \frac{Q}{6(H_1 - H_2 c^2)}, \qquad (6.33)$$

where $a_\pm$ are the roots of a quadratic equation

$$QV^2/6 + PV/3 + (1 - c^2) = 0 \qquad (6.34)$$

and a change of variable $V = 1/y$ is used:

$$\left(-\frac{1}{y^2}y'\right)^2 = \frac{1}{y^2}\left(\frac{1}{y} - a_+\right)\left(\frac{1}{y} - a_-\right)\frac{Q}{6(H_1 - H_2 c^2)}. \qquad (6.35)$$

Then Eq. (6.35) is multiplied by y^4 and a series of straightforward algebraic operations are carried out [44], arriving to

$$y' = \pm\sqrt{\frac{Qa_+a_-}{6(H_1 - H_2c^2)}}\sqrt{(y - a)^2 - b^2},$$ (6.36)

where $a = (a_+ + a_-)/2a_+a_-$ and $b = (a_+ - a_-)/2a_+a_-$ have been introduced for convenience. Since $y' = dy/d\xi$, then after integration we get

$$\sqrt{\frac{Qa_+a_-}{6(H_1 - H_2c^2)}}\,\xi = \pm\,\text{arccosh}\left(\frac{y - a}{b}\right).$$ (6.37)

Solving this equation for y and then using $V = 1/y$, the following solutions are obtained:

$$U_1(\xi) = \frac{-6(1 - c^2)}{P + P\sqrt{1 - 6(1 - c^2)Q/P^2}\,\cosh(\xi\sqrt{(1 - c^2)/(H_1 - H_2c^2)})},$$ (6.38a)

$$U_2(\xi) = \frac{-6(1 - c^2)}{P - P\sqrt{1 - 6(1 - c^2)Q/P^2}\,\cosh(\xi\sqrt{(1 - c^2)/(H_1 - H_2c^2)})}.$$ (6.38b)

It is easy to check that these expressions are indeed solutions to Eq. (6.30) by making a substitution $\xi = X - cT$ and then plugging expressions (6.38) into Eq. (6.30). Solutions (6.38) are demonstrated in Fig. 6.7.

6.2.3 Solution Types

Expressions (6.38) represent constant profile solutions for Eq. (6.30). These solutions are exact and can represent either solitary or periodic wave solutions. The nature of a solution depends on the coefficients P, Q, H_1, H_2 and c [21, 43, 44]. In this subsection the solutions (6.38) are discussed.

The existence of solitary and periodic solutions

The most obvious limitation for the coefficients P, Q and c comes from the square root in front of the hyperbolic cosine – the real solutions of expressions (6.38) are only possible in case when the following condition is satisfied:

$$\left(1 - c^2\right)\frac{Q}{P^2} < \frac{1}{6}.$$ (6.39)

This condition sets general limits to velocity c:

$$c > \sqrt{1 - \frac{P^2}{6Q}} \quad \text{for} \quad Q > 0 \quad \text{and} \quad c < \sqrt{1 - \frac{P^2}{6Q}} \quad \text{for} \quad Q < 0.$$ (6.40)

Fig. 6.7 Solutions of
Eq. (6.30) in case of (top
panel) $P = -10$, $Q = 40$
and (middle, bottom panels)
$P = -8$, $Q = -130$. For all
profiles $c = 0.8$, $H_1 = 4$ and
$H_2 = 5$. In case of $Q > 0$ only
solution (6.38a) exists (top
panel) and in case of $Q < 0$
solutions (6.38a) and (6.38b)
coexist (middle, bottom panels). The moving coordinate
ξ is on the horizontal and
the amplitude $U(\xi)$ is on
the vertical axis. Reproduced
with permission from [44];
©Elsevier 2019.

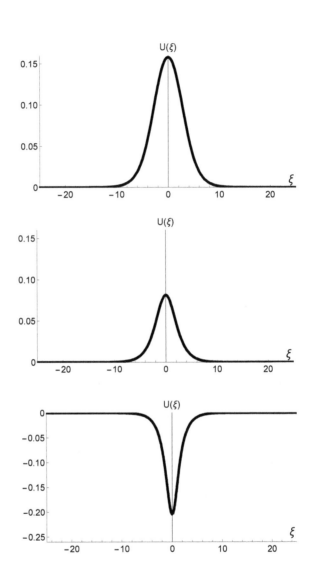

Depending on the choice of parameters P, Q, H_1, H_2 and c, additional restrictions
on the velocities may emerge. Note that it is experimentally established that for a
biomembrane in a fluid state $P < 0$ and $Q > 0$ [26] and in gel state $P > 0$ and $Q > 0$
[45].

In addition, if solitary wave solutions are desired, then a condition

$$\frac{1 - c^2}{H_1 - H_2 c^2} > 0 \qquad (6.41)$$

inside square root in hyperbolic cosine must be satisfied. In case of the negative root, periodic solutions emerge since $\cosh(ix) = \cos(x)$. In case of $H_2 = 0$, as it is the case in Eq. (6.28), solitary wave solutions emerge in case of $c < 1$ and periodic waves in case of $c > 1$ provided that condition (6.39) is satisfied.

Phase analysis

The classical approach for analysis of the existence of solitary waves is based on the polynomial at the rhs of Eq. (6.32), which is also known as the 'pseudopotential' [1, 9, 11]:

$$\Phi_{eff}(V) = (1 - c^2)V^2 + \frac{1}{3}PV^3 + \frac{1}{6}QV^4. \qquad (6.42)$$

Solitary wave solutions exist when there is a double zero, zeros are real, $\Phi_{eff}(V) > 0$ and there is a local maximum next to the double zero [1, 21].

For Eq. (6.30) the four zeros of the polynomial (6.42) are

$$V_{1,2} = 0 \quad \text{and} \quad V_{3,4} = P/Q\left(-1 \pm \sqrt{1 - 6(1 - c^2)Q/P^2}\right). \qquad (6.43)$$

Since the value of $V_{1,2}$ does not depend on the choice of coefficients, then the requirement for a double zero is always fulfilled at the origin. The values of zeros V_3 and V_4 depend on the choice of parameters – in case of $Q > 0$ the zeros of the polynomial (6.42) have the same sign and there will be only one bounded region where $\Phi_{eff}(V) > 0$. In case of $Q < 0$ the zeros V_3 and V_4 have opposite signs and consequently two regions with $\Phi_{eff}(V) > 0$ exist. This is demonstrated in Fig. 6.8 for the case of $P < 0$. For the case of $P > 0$ the zeros V_3 and V_4 have opposite signs and the shape of the 'pseudopotential' is flipped with respect to the vertical axis [21, 43, 44].

The dynamic behaviour of complicated nonlinear ODEs can be understood through the analysis of phase portraits. In addition to the existence of solitary wave solutions, this method gives insight to the existence of other kind of solutions. To that end, Eq. (6.31) is rewritten as a system of first order ODEs:

$$V' = W, \qquad (6.44a)$$

$$W' = (H_1 - H_2 c^2)^{-1}\left[(1 - c^2)V + \frac{1}{2}PV^2 + \frac{1}{3}QV^3\right]. \qquad (6.44b)$$

Fixed points V^* are found by setting $V' = W' = 0$:

$$V_1^* = 0, \quad V_{2,3}^* = \frac{3}{4}\frac{P}{Q}\left(-1 \pm \sqrt{1 - \frac{16}{3}(1 - c^2)Q/P^2}\right) \qquad (6.45)$$

Fig. 6.8 Shape of the 'pseudopotential' (6.42) in case
of $Q > 0$ (top panel) and in
case of $Q < 0$ (bottom panel).
Coefficients are same as in
Fig. 6.7: (top panel) $Q = 40$;
(bottom panel) $Q = -130$.
Reproduced with permission
from [44]; ©Elsevier 2019.

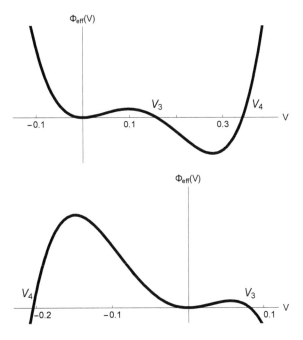

and the nature of the fixed points is found by finding the eigenvalues λ of the Jacobian
matrix for system (6.44) for each fixed point V^* [44, 54]:

$$\lambda_{1,2} = \pm i \sqrt{\frac{-PV - QV^2 - (1 - c^2)}{H_1 - H_2 c^2}}. \tag{6.46}$$

It can be seen in Eqs. (6.45) and (6.46) that while the coordinates of the fixed points
depend on the nonlinear coefficients P, Q and velocity c, the nature of the fixed points
depends on the same coefficients and the dispersion type. In case of Eq. (6.30) the
fixed points are either a saddle (real eigenvalues) or a centre (imaginary eigenvalues)
depending on the choice of coefficients [21, 44]. For the existence of solitary wave
solutions there has to be a saddle point at the double zero and a homoclinic orbit.

The case of $H_1 - H_2 c^2 > 0$

Expressions (6.38) can represent either solitary or periodic wave solutions. The exact
nature of these solutions depends on the choice of parameters P, Q, H_1, H_2 and c.
When condition $H_1 - H_2 c^2 > 0$ is fulfilled, then solitary wave solutions exist in case
of $c < 1$. In case of $c > 1$ the root inside hyperbolic cosine becomes imaginary
and solitary wave solutions are not possible. This can also be deduced with phase
analysis – although the double zero will always remain at the origin (see Eq. (6.43)),
the fixed point V_1^* at the origin will become a centre (see Eq. (6.46)).

The effect of the coefficient Q on the solutions is significant. In case of $Q > 0$, which is the case of a biomembrane, only solution (6.38a) represents solitary wave. This is also demonstrated in Fig. 6.9 where it can be seen that in case of $Q > 0$ one homoclinic orbit exists – in case of $P < 0$ on the right side and in case $P > 0$ on the left side of vertical axis.

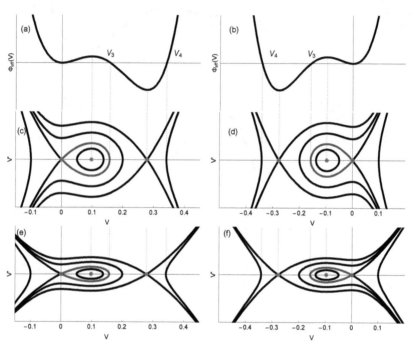

Fig. 6.9 Shape of the 'pseudopotential' (6.42) – at top panels and phase portraits of Eq. (6.32) in case of $Q > 0$. Here $c = 0.8$, $Q = 40$, $|P| = 10$, $H_1 = 4$, while $H_2 = 5$ for the middle panels and $H_2 = 0$ for the bottom panels. Homoclinic orbit representing solution (6.38a) is shown in blue; fixed points (6.45) are shown in red. Adapted from [21].

In case of $Q < 0$, two homoclinic orbits exist and both solutions (6.38) represent solitary wave solutions. As in case of $Q > 0$, the effect of coefficient P is on the polarity of solutions. This is demonstrated in Fig. 6.10.

The simultaneous existence of two homoclinic orbits can also give rise to soliton doublet – coexisting solitary waves with same velocities but different amplitudes. This is demonstrated in Fig. 6.11. The solution in Fig. 6.11 is obtained by integrating system (6.44) by an ODE solver [59] with a suitable initial amplitude – either V_3 or V_4 (see Eq. (6.43)). We seek for a solution which is slightly out of the homoclinic orbit. The solution demonstrating 'nearly solitary' waves [48] which may be called a 'soliton doublet'. Note that in calculations, the doublet will be repeated after long time. The co-existing solitons are described by expressions (6.38a) and (6.38b) where the free parameters c, P and Q determine the amplitude of solitons (cf. with the classical soliton [1]).

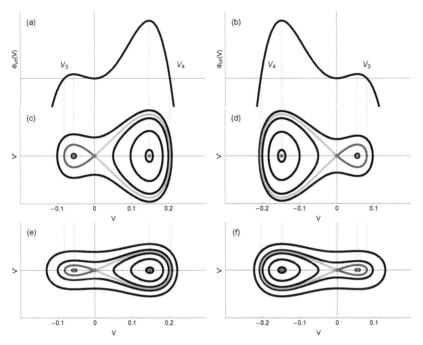

Fig. 6.10 Shape of the 'pseudopotential' (6.42) – at top panels and phase portraits of Eq. (6.32) in case of $Q < 0$. Here $c = 0.8$, $Q = -130$, $|P| = 8$, $H_1 = 4$, while $H_2 = 5$ for the middle panels and $H_2 = 0$ for the bottom panels. Homoclinic orbit for solutions (6.38a) and (6.38b) are shown in blue and green respectively; fixed points (6.45) are shown in red. Adapted from [21].

Fig. 6.11 The numerical solution of Eq. (6.44) demonstrating the soliton doublet corresponding to the phase space depicted in Fig. 6.10. Here $c = 0.8$, $P = -8$, $Q = -130$, $H_1 = 4$ and $H_2 = 5$. Reproduced with permission from [44]; ©Elsevier 2019.

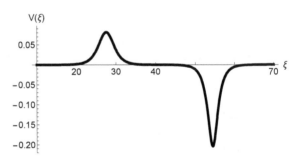

It can also be shown that soliton doublet will propagate at constant velocity while maintaining its shape. The solution is obtained numerically by making use of the Discrete Fourier Transform (DFT) based pseudospectral method (PSM) [21, 50]. This is demonstrated in Fig. 6.12. The existence of a soliton doublet for the HJ model is mentioned also by Vargas et al. [58].

In case of $Q < 0$ and $1 < c < \sqrt{1 - P^2/6Q}$ expressions (6.38) represent periodic wave solutions. It can be shown that in this case, 'pseudopotential' (6.42) is positive between the points V_3 and V_4 and a stable orbit exists (shown in blue) which means existence of a periodic solution (see Fig. 6.13). The transition from solitary wave solutions to a periodic solution can also be seen in solution (6.38) since the imaginary root under hyperbolic cosine follows the aforementioned condition. In this case solutions (6.38a) and (6.38b) represent identical coexisting periodic waves with a phase difference of π (see Fig. 6.13). What is interesting is that the phase portrait, in this case, looks similar to Fig. 6.10. However, it has been slightly shifted to the right and the higher amplitude part is realised. A solution to the smaller homoclinic orbit can be obtained numerically. This is demonstrated in Fig. 6.14. The existence of periodic solutions for the HJ model is shown also by Vargas et al. [58].

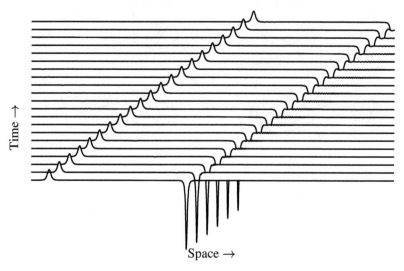

Fig. 6.12 A soliton doublet: time slice plot. Here $c = 0.99$, $P = -0.5$, $Q = -1.7946877$, $H_1 = 0.5$, $H_2 = 0.25$. The total length of the space is 512π and a segment from 128π to 512π is shown and $\Delta T = 30$. Reproduced with permission from [44]; ©Elsevier 2019.

In case of $Q > 0$ and $c > 1$ there exists one focus and two unstable nodes. Also, pseudopotential $\Phi_{eff}(V)$ has negative values between the zeros and expressions (6.38) do represent travelling wave solutions.

It can also be seen in Figs. 6.9 and 6.10 that in the case of the second dispersion coefficient H_2 a more localised solution is obtained: the greater value of quantity V' means the steeper slope (and hence the smaller width) of the solitary wave. The effect of dispersive term H_2 on the width of a solitary wave is demonstrated in Fig. 6.15, where it can be seen that higher values of H_2 result in more localised solutions. This effect is important for experimental studies.

The case of $H_1 - H_2c^2 < 0$

Mathematically it is also possible to obtain solutions for the case of $H_1 - H_2c^2 < 0$. Since Eq. (6.32) can be thought of as a conservation of 'pseudoenergy', then if

Fig. 6.13 Solution (6.38) may also represent two periodic waves with a phase difference of π when $(1 - c^2)/(H_1 - H_2 c^2) < 0$. Here at top panel – phase portrait, at bottom panel – wave profiles. Solution (6.38a) is shown in solid blue and solution (6.38b) in dashed red. Here $c = 1.2$, $P = 18$, $Q = -80$, $H_1 = 2$ and $H_2 = 1$. Reproduced with permission from [44]; ©Elsevier 2019.

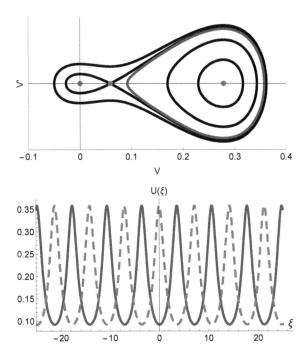

Fig. 6.14 Fixed point at the origin (V_1^*) becomes a centre in case of $(1 - c^2)/(H_1 - H_2 c^2) < 0$ and periodic solutions around it can be obtained numerically. Here at top panel – phase portrait, at bottom panel – wave profiles. Coefficients are same as in Fig. 6.13. Reproduced with permission from [44]; ©Elsevier 2019.

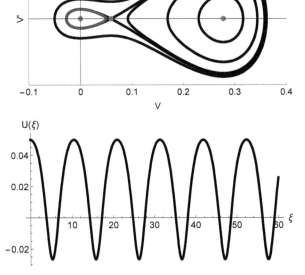

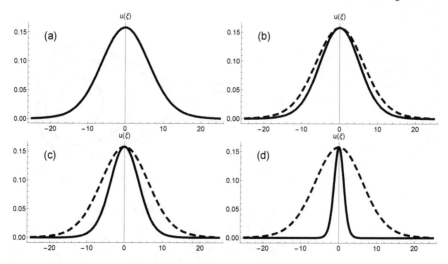

Fig. 6.15 The effect of the second dispersive term $H_2 U_{XXTT}$ on the width of a solitary wave. Here $c = 0.8$, $P = -10$, $Q = 40$, $H_1 = 4$; (a)$H_2 = 0$, (b)$H_2 = 2$, (c)$H_2 = 4$ and (d)$H_2 = 6$. Reproduced with permission from [21]; ©Taylor & Francis 2017.

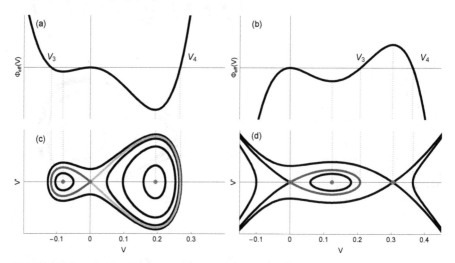

Fig. 6.16 Existence of solitary waves in case of $H_1 - H_2 c^2 < 0$ and $c > 1$. Here $c = 1.1$, $Q = 40$, $P = -3$, $H_1 = 4$ and $H_2 = 5$ (left column) and $c = 1.2$, $Q = -35$, $P = 10$, $H_1 = 4$ and $H_2 = 5$ (right column). 'Pseudopotential' shown in top panels, phase portrait – in bottom panels. The homoclinic orbit for solution (6.38a) is shown in blue and for (6.38b) is shown in green; fixed points (6.45) are shown in red. Adapted from [21].

condition $H_1 - H_2c^2 < 0$ is satisfied then Φ_{eff} has also to be negative. However, in this case the behaviour of the solutions is reversed – solitary wave solutions emerge when $c > 1$ and periodic – when $c < 1$. Also, the effect of coefficient Q to solutions is reversed – in case of $Q > 0$ two homoclinic orbits exist and in case of $Q < 0$ only one homoclinic orbit exists. An example is shown in Fig. 6.16.

6.2.4 Solutions Emerging from Arbitrary Initial Conditions

The influence of coefficients P, Q, H_1 and H_2 on the evolution of solutions can be demonstrated with solving the Eq. (6.30) under localised initial condition making use of the pseudospectral (PSM) method. The advantages and disadvantages of the PSM are well explored in the literature [24, 25].

Here two points are worth to be stressed: (i) the PSM requires using periodic boundary conditions, (ii) the governing equations have to be in a suitable form for applying the PSM with time derivatives on the lhs and spatial derivatives on the rhs of the equation. The first point is not a problem, however, taking a look at Eq. (6.30) it is evident that we have a mixed partial derivative term U_{XXTT}. Consequently, a change of variables is used for transforming the governing equation (6.30) for allowing the application of the PSM [19, 40, 55]. As shown, the basic idea of the PSM is to find the spatial derivatives by making use of the properties of the Fourier transform and then solve the resulting ODE with respect to time derivative using available schemes for numerical solving of the ODE's (see Appendix A).

For numerical analysis a following pulse-type localised initial condition is used:

$$U(X, 0) = U_o \operatorname{sech}^2 B_o X, \tag{6.47}$$

where $U_o = 1$ and $B_o = 1$ are the amplitude and the width of the pulse, respectively. As required by the PSM, the system is solved under periodic boundary conditions:

$$U(X, T) = U(X + 2km\pi, T), \tag{6.48}$$

where $m = 1, 2, \ldots$ and $k = 12$, which means that the total length of the spatial period is 24π. The initial condition for the velocity is $U(X, 0)_T = 0$ meaning that the initial pulse splits into two counterpropagating pulses like in case of the classical wave equation. Some examples are shown using different combinations of parameters in which case the used parameters are noted separately. Although the initial condition is strictly speaking not a periodic function and first derivatives involve discontinuities across the boundary point, the numerical error using the PSM with such a periodic boundary condition is small. This is clearly demonstrated by the detailed analysis of the applicability of the PSM [50].

In addition to the formation of solitary waves, it is possible that several different waveprofile regimes exist for the solutions of the governing equations (6.30) depending on the parameters but also on the initial conditions. To name the ones investigated previously:

(i) Airy or reverse Airy-type oscillatory structures (see Fig. 6.17);
(ii) solitary waves; single or as a part of solitary wave train (see Figs. 6.17, 6.18);
(iii) a hybrid solution where part of the initial pulse evolves into a train of solitary waves and remainder of the initial pulse forms an oscillatory structure [40, 42, 55].

Fig. 6.17 Waveprofile plots for the normal (top panel, $T = 1500$) and anomalous (bottom panel, $T = 1700$) dispersion cases for the positive (solid black line) and negative (blue dotted line) initial condition amplitudes. Waveprofile propagation direction is from left to right. Parameters: $U_o = \pm 1$, $B_o = 1/8$, $k = 128$, $n = 1024$, $c = 1$, $P = -0.1$, $Q = 0.01$, $H_2 = 0.5$ and $H_1 = 0.28125$ (normal dispersion), $H_1 = 0.78125$ (anomalous dispersion). Reproduced with permission from [21]; ©Taylor & Francis 2017.

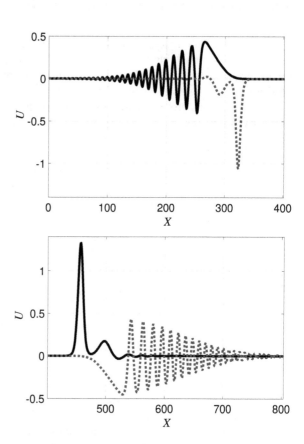

From the viewpoint of nerve pulse propagation the most interesting solution is the solitary wave, however, the rest of the solution types can not be ignored either as these might somehow be relevant for either nerve pulse propagation or some kind of pathologies or for cell deformation in general. It should be stressed that not only the equation parameters are important in determining what kind of solution evolves from the initial excitation but also the character of an initial excitation is important. An example is shown in Fig. 6.17 where some solutions corresponding to the different parameters and initial condition amplitudes are presented. Depending on the dispersion type, the sign of the initial excitation plays also a significant

Fig. 6.18 Waveprofiles comparison plot (top panel) at $T = 1750$ for the negative (blue dotted line) against positive (black solid line) initial condition amplitude. Lower amplitude solitary waves are propagating faster. Direction of propagation from left to right. Parameters: $P = -0.1$, $Q = 0.05$, $H_1 = 0.5$, $H_2 = 0.5$, $k = 128$, $n = 1024$, $U_0 = \pm 1$, $B_0 = 1/8$, $c = 1$, $T = 0 \ldots 1750$. Example of evolved solitary wave train is shown in the bottom panel [55]. Reproduced with permission from [21]; ©Taylor & Francis 2017.

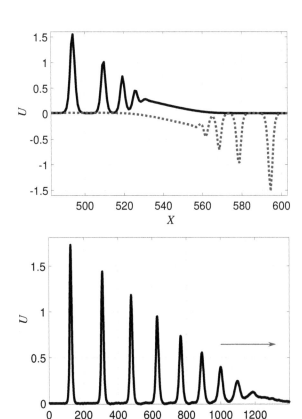

role – namely it determines whether the emerging wave structure is composed of solitary pulses or Airy or reverse-Airy type oscillatory packet under the parameter combination used in Fig. 6.17. Another extremely interesting phenomenon to be mentioned is the case where smaller amplitude solitary waves travel faster than the high amplitude ones shown in Fig. 6.18. Under the suitable parameter combinations it is possible to demonstrate that both negative and positive amplitude solitary waves can exist simultaneously. Moreover, if the smaller amplitude solitary waves travel faster then the larger negative amplitude solitary waves travel even faster. However, this is not a universal symmetry but depends on the right ratio of parameters. The most common solution type seems to be an oscillatory structure with a few solitary pulses. This is known in soliton dynamics [9, 11] by noting that some part of the initial pulse energy is sufficient to form one or more solitary pulses and the remainder forms an oscillatory trail either in front or behind (depending on the dispersion type) of the propagating solitary waves.

Also, we track the waveprofile peak trajectories by finding the exact local maxima of the wave profiles. This is done by making use of the properties of the Fourier transform [50] for finding the exact spatial coordinates of the pulse peaks at each time step to find the waveprofile velocities (reconstructing the wave profile from the Fourier spectrum to minimise inaccuracies from using the discrete grid).

Some observations follow from such an analysis:

(i) as expected, the dispersion parameters H_1 and H_2 have a strong effect on the evolution of the wave profiles. It means that the main pulse velocities are clearly different depending on the dispersion parameters and in addition, the dispersion type determines on which side (relative to the propagation direction) the secondary wave structures emerge from the main pulse. In physical terms, increasing the parameter H_1 increases the main pulse propagation velocity as predicted by dispersion analysis [40];

(ii) The nonlinear parameters P and Q have also certain influence on the waveprofile propagation velocities. In the case of the normal dispersion ($c_1/c_0 < 1$) increasing the nonlinearity means the slower propagation velocity for the wave profiles. In the case of the anomalous dispersion ($c_1/c_0 > 1$), the main pulse velocity remains almost the same. However, in physical terms, the effect is more significant for the secondary oscillatory structures meaning that in the case of higher nonlinear parameters the secondary structures have propagated at higher velocity. This is in agreement with previous results [55]. Namely, it has been demonstrated that due to the uncommon (in the context of Boussinesq-type equation) nonlinear terms certain parameter combinations lead to the situation where the smaller amplitude solitary waves propagate faster than the higher amplitude ones [55].

Let us take a more detailed look at how the nonlinear and dispersive parameters influence the observable quantities of the wave profiles under the initial conditions which differ in sign. The speed of the peak of the main pulse can be observed by tracking the coordinates of the peak of the main pulse. It is done by reconstructing the waveprofile shape from the full Fourier spectrum [50] at each time step. In the further analysis, parameters P and Q are changed from -0.9 to $+0.9$ with the step size of 0.1 and for dispersion related parameters H_1 and H_2 three combinations are used for calculation: a normal dispersion case ($H_1 = 0.3$, $H_2 = 0.7$), a 'balanced dispersion' case ($H_1 = H_2 = 0.5$) and an anomalous dispersion case ($H_1 = 0.7$, $H_2 = 0.3$).

The notation of 'balanced dispersion' needs some clarification as it is not in common use compared with anomalous and normal dispersion types. In our context, it is a situation where two dispersion terms in the governing equation have opposite signs and their influence is balanced. It must be stressed that this is not a dispersionless case because it still leads to slight (almost unnoticeable) dispersion in wave profiles.

As stated by Heimburg and Jackson [26], for biomembranes the nonlinear parameters fulfil the conditions $P < 0, Q > 0$.

The case $P < 0, Q > 0$ leads to the following conclusions:

Normal dispersion case. The initial condition with a negative amplitude leads to a greater main pulse velocity than the initial condition with a positive amplitude. Both cases demonstrate that decreasing the nonlinear parameter Q (towards the zero) leads to a small decrease of the main pulse velocity. In the case of the initial condition with a negative amplitude, the main pulse amplitude is greater than in the case of the initial condition with a positive amplitude and the observed oscillations are larger for the case with positive initial amplitude than in the case with the negative initial amplitude. Increasing parameter P leads to decrease in the main pulse velocity in the case of the negative amplitude initial condition while in the case of the positive initial amplitude the main pulse velocity remains the same. Increasing parameter P towards zero leads to marginally greater amplitude for the main pulse in the case of the initial condition with a negative amplitude while in the case of the initial amplitude with a positive amplitude the main pulse amplitude is unaffected by the changes in the nonlinear parameter P. The magnitude of the oscillatory structure is unaffected in the normal dispersion case.

Anomalous dispersion case. The main pulses propagate with a velocity greater than one (it is a dimensionless case) under both of the initial condition signs. However, the main pulse amplitudes and associated oscillatory structures are different. The changes are the following. Increasing parameter Q leaves the observed propagation speed the same but decreases the observed main pulse amplitude and leaves the observed oscillatory structures unchanged. Increasing the parameter P does not affect the main pulse velocity significantly in the considered dispersion case regardless of the sign of the initial amplitude. However, by increasing nonlinear parameter P, the main pulse amplitude will be decreased and the amplitudes of the oscillatory structures will be increased under both considered signs of the initial condition. Some more cases are also analysed in [20].

6.2.5 Interaction of Solitons

In general, a soliton is defined as a stable particle-like state of a nonlinear system [10]. Another way of describing the phenomenon called soliton is through its properties. A soliton is a wave in the nonlinear environment that (i) is localised in space, (ii) has a stable form and (iii) restores its speed and structure after interaction with another soliton [11, 15]. Solitons emerge when there is a balance in the physical system between dispersive and nonlinear effects. In essence, it can be said that solitons are nonlinear waves that behave between interactions like linear waves propagating without any change of their shape. A solitary wave is a wave in the nonlinear environment where the key properties of solitons are not all strictly fulfilled. For example, if the interaction between two waves is not entirely elastic (or it is not possible to observe the interaction) or if the form of the wave is not sufficiently stable in time, then the wave is often called a solitary wave to distinguish it from the soliton. This is a case in many physical systems.

In Fig. 6.19 one can see the solitary wave propagation (top panel) and interaction (bottom panel) modelled by the iHJ equation (6.30). The parameters for calculations are the same as in Sect. 6.2.3 except $H_2 = 0$. From Fig. 6.19 it is clear that while the single iHJ pulse is stable, it is strictly speaking a solitary wave, not a soliton because the interaction with another such wave is not elastic. There is significant

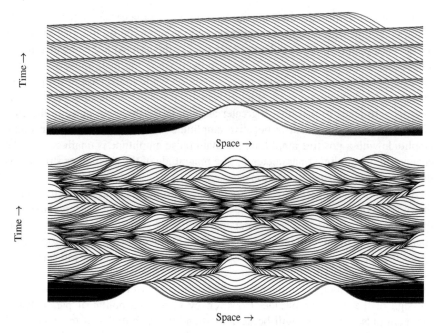

Space →

Space →

Fig. 6.19 The time slice plots for the case $H_2 = 0$ (Eq. (6.30)). The solitary wave solution (top panel) and interaction of counterpropagating solitary waves (bottom panel). Reproduced with permission from [21]; ©Taylor & Francis 2017.

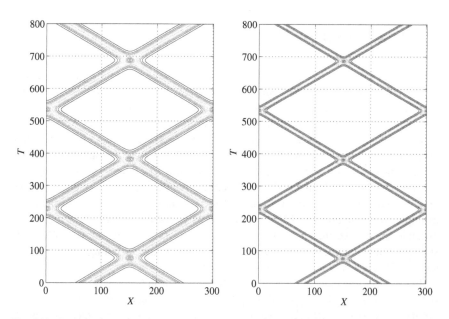

Fig. 6.20 Contour plots of interactions of solitonic solutions. Parameters $c = \pm0.99$, $P = -10$, $Q = 40$, $H_1 = 1$, $H_2 = 0$ (left panel) and $H_2 = 0.75$ (right panel). Reproduced with permission from [21]; ©Taylor & Francis 2017.

Fig. 6.21 Waveprofile plots (top panel) and corresponding phase plots (bottom panel) at $T = 770$ after five interactions. Parameters $c = \pm 0.99$, $P = -10$, $Q = 40$, $H_1 = 1$. Only a waveprofile propagating to the left is shown. Reproduced with permission from [21]; ©Taylor & Francis 2017.

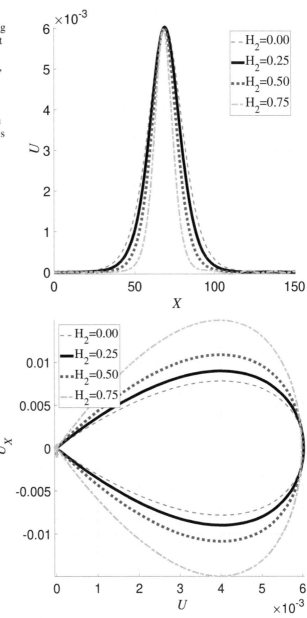

radiation even during the first interaction event and the shape of the waveprofile is
not properly restored after the interaction. However, it should be noted that some
parameter combinations can exist which result in relatively stable solutions with
almost no radiation.

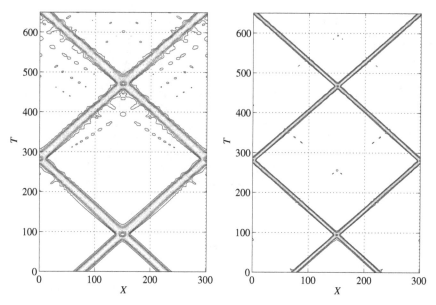

Fig. 6.22 Contour plots of interactions of solitonic solutions. Parameters $c = \pm 0.8$, $P = -10$,
$Q = 40$, $H_1 = 4$, $H_2 = 0$ (left panel) and $H_2 = 5$ (right panel). Reproduced with permission from
[21]; ©Taylor & Francis 2017.

The 'time slice' plot shown in Fig. 6.19 is useful for giving an overview 'at a
glance' of the evolution of a solution in time. Spatial coordinate is on the horizontal
and time on the vertical axis. Due to the periodic boundary conditions, profiles
moving out of the frame on the left enter the frame on the right.

The character of the interaction of single solitary waves depends on the parameters
of the model and consequently, on the dispersion type (normal, anomalous). Here the
following set of parameters is used: $c = \pm 0.99$, $P = -10$, $Q = 40$, $H_1 = 1$, $H_2 = 0$,
$H_2 = 0.25$, $H_2 = 0.50$ and $H_2 = 0.75$. In Fig. 6.20 one can see the interactions
when the parameter $H_2 = 0$ (left panel) and when $H_2 = 0.75$ (right panel) – are
remarkably similar and non disruptive. The main difference between the examples
is that the solitonic waveprofiles are more localised for the case $H_2 \neq 0$. In this
case, the interactions have almost no radiation (negligible radiation two orders of
magnitude smaller than the main pulse amplitude). Amplitude isolines in Fig. 6.20
are separated by 0.001 from 0.001 to 0.013.

In Fig. 6.21 the waveprofiles and the corresponding phase plots are presented
after the five interaction events ($T > 700$). The solitonic waveprofiles corresponding
to higher values of H_2 are more localised as expected (the waveprofile in Fig. 6.21

is propagating to the left). The small distortions to the waveprofiles are easier to recognise in phase plots (bottom panel), in particular the small radiation close to zero which is noted being two orders of magnitude smaller than the main pulse under the used parameter combination.

Let us return to a parameter set presented in Sect. 6.2.3 for the analytical solution of the model iHJ equation in the form of a soliton. It turns out that there is also a possible scenario where the solitonic solutions with the additional dispersive term are more stable through interactions than the solitonic solutions involving parameter $H_2 = 0$. In Fig. 6.22 the case $H_2 = 0$ is presented at the left panel and the case $H_2 = 5$ at the right panel. The amplitude isolines are separated by 0.02 from 0.02 to 0.3. It is clear that under the parameter set used in Figs. 6.19 and 6.22 the solitonic waves corresponding to $H_2 = 0$ have more radiation than the case $H_2 = 5$ which is relatively stable in comparison throughout interactions. Neither of the cases can be considered solitons in the strict mathematical sense [11] as in both cases there is the significant enough radiation after only three interactions.

6.2.6 Discussion

The systematic analysis of solutions to the special Boussinesq-type equation with the displacement-dependent nonlinearities has revealed several interesting phenomena [20, 21, 44]. The analysis presented above is focused on Eq. (6.29) (or its dimensionless form (6.30)) which is the improved Heimburg-Jackson (iHJ) model for describing the longitudinal wave process in biomembranes. Like every wave equation, it describes the process generated by initial and/or boundary conditions expressed in terms of the dependent variable. Here the variable under consideration is the change of the density in the longitudinal direction that can also be described as the longitudinal displacement. It must be stressed that the conventional wave propagation theory in continuum mechanics [7] involves deformation-dependent nonlinearities while here the governing nonlinear wave equation involves displacement-dependent nonlinearities. In terms of Eq. (6.30), the existence of solitary solutions is demonstrated, the emergence of trains of solitary pulses is shown and the properties of emergence analysed, and the interaction of single solitary waves and trains studied.

The analysis can be summarised with the following conclusions:

- The additional dispersive term U_{XXTT} with coefficient H_2 in the dimensionless form in addition to the *ad hoc* dispersive term U_{XXXX} [26] describes actually the influence of the inertia of the microconstituents (lipids) of the biomembrane. This corresponds to the understandings of continuum mechanics of microstructured solids [36] and is demonstrated also experimentally [35]. Term U_{XXTT} regulates the width of the solitary pulse (see Fig. 6.15) and such an effect can be used for determining the value of H_2 (and correspondingly h_2) from experiments. It also determines how fast the transition from the low frequency speeds to the high frequency speeds occurs (see Fig. 6.6);

- The improved model (6.30) removes the discrepancy that at higher frequencies the velocities are unbounded (see Fig. 6.6);
- The fourth-order pseudopotential (6.42) derived from equation (6.30) involves several solution types of solitary waves and under certain conditions ($Q > 0$, $H_2 >> H_1$) an oscillatory solution exists (see Figs. 6.13 and 6.14);
- As expected, soliton trains emerge from an arbitrary initial condition. These results were obtained as a result of the numerical simulation by using the pseudospectral method [50]. It has been shown that the nonlinear effects start to influence the emergence either from the front or from the back of the propagating pulse (see Fig. 6.18) depending on the signs of coefficients Q and P. For the case of a biomembrane [26], the condition $Q < 0$, $P > 0$ and the train emerging from a positive input starts with smaller solitons which travel faster than the bigger ones. This is different from the conventional case of nonlinear evolution equations (the KdV equation, for example). Such a result is extremely important from a general viewpoint of soliton dynamics. In the case of negative input, the train is headed by bigger solitons which travel faster (see Fig. 6.18). It has been shown that there are several wave types possible: solitary waves (Fig. 6.17), oscillatory (Airy-type) waves (Fig. 6.17), and hybrid solutions.
- Like in many real physical systems, the interaction of solitary waves is not fully elastic (see Figs. 6.20, 6.22) demonstrating that these solitary waves are not solitons in the strict sense [11]. However, like in other Boussinesq-type equations [7, 18], the radiation effects accompanying every interaction start cumulating rather slowly and the interacting solitons keep their shape for a rather long time. It gives the ground to call emerging solitary waves modelled by Eq. (6.30) solitons like it is done in other physical cases [33].

Biological structures, as a rule, have high complexity over many scales because the macrobehaviour is strongly influenced by the embedded microbehaviour. Mathematical modelling is a tool not only for describing biological processes and performing experiments *in silico* but to understand the process. The behaviour of a biomembrane is an excellent example of how the microstructure (lipids) of a membrane has a direct impact on wave phenomena along the membrane. The analysis of the governing equation (6.30) presented above demonstrates the richness of the model from the viewpoint of mathematical physics and opens the ways for further physiological experiments concerning the properties of biomembranes.

6.3 The Wave Equations with Coupling

In case of electroelasticity, the elastic and electric effects should be coupled. For example, in elastic ferroelectrics the longitudinal displacement (the 1D setting) and the rotation of dipoles can be described by a system of two wave equations with coupling forces [32]:

$$U_{tt} - c_L^2 U_{xx} = -\frac{1}{2}\alpha \left(\Phi_x^2\right)_x, \tag{6.49}$$

$$\Phi_{tt} - \Phi_{xx} - \Psi \sin \Phi = -\alpha (U_x \Phi_x)_x, \tag{6.50}$$

where U is the longitudinal displacement, Φ is the rotation angle, c_L is the velocity, α is the piezoelectric coefficient and Ψ is the electric susceptibility (original notations are used). Equation (6.49) is a wave equation with forcing, Eq. (6.50) is a Sine-Gordon equation with forcing. System (6.49)-(6.50) is an excellent example to demonstrate how the physical effects are coupled by reciprocal forces.

References

1. Ablowitz, M.J.: Nonlinear Dispersive Waves. Asymptotic Analysis and Solitons. Cambridge University Press, Cambridge (2011). DOI 10.1017/CBO9780511998324
2. Andronov, A., Witt, A., Khaikin, S.: Theory of Oscillations. Phys. Math. Publ., Moscow (In Russian) (1959)
3. Appali, R., Petersen, S., van Rienen, U.: A comparision of Hodgkin-Huxley and soliton neural theories. Adv. Radio Sci. **8**, 75–79 (2010). DOI 10.5194/ars-8-75-2010
4. Billingham, J., King, A.C.: Wave Motion. Cambridge Texts in Applied Mathematics. Cambridge University Press (2001). DOI 10.1017/CBO9780511841033
5. Bonhoeffer, K.F.: Acativation of passive iron as a model for the excitation of nerve. J. Gen. Physiol. **32**(1), 69–91 (1948). DOI 10.1085/jgp.32.1.69
6. Bountis, T., Starmer, C.F., Bezerianos, A.: Stationary pulses and wave front formation in an excitable medium. Prog. Theor. Phys. Suppl. **139**(139), 12–33 (2000). DOI 10.1143/PTPS.139.12.
7. Christov, C.I., Maugin, G.A., Porubov, A.V.: On Boussinesq's paradigm in nonlinear wave propagation. Comptes Rendus Mécanique **335**(9–10), 521–535 (2007). DOI 10.1016/j.crme.2007.08.006
8. Courtemanche, M., Ramirez, R.J., Nattel, S.: Ionic mechanisms underlying human atrial action potential properties : insights from a mathematical model. Am. J. Physiol. **275**(1), 301–321 (1998)
9. Dauxois, T., Peyrard, M.: Physics of Solitons. Cambridge University Press, Cambridge (2006)
10. Dodd, R., Eilbeck, J., Gibbon, J., Morris, H.: Solitons and nonlinear wave equations. Academic Press Inc. LTD., London (1982)
11. Drazin, P., Johnson, R.: Solitons: an Introduction. Cambridge University Press, Cambridge (1989)
12. Engelbrecht, J.: On theory of pulse transmission in a nerve fibre. Proc. R. Soc. A Math. Phys. Eng. Sci. **375**(1761), 195–209 (1981). DOI 10.1098/rspa.1981.0047.
13. Engelbrecht, J.: Nonlinear Wave Processes of Deformation in Solids. Pitman Advanced Publishing Program, London (1983)
14. Engelbrecht, J.: Introduction to Asymmetric Solitary Waves. Longman Scientific & Technical and Wiley, New York (1991)
15. Engelbrecht, J.: Beautiful dynamics. Proc. Estonian Acad. Sci. Physics/Mathematics **1**(44), 108–119 (1995)
16. Engelbrecht, J.: Questions About Elastic Waves. Springer International Publishing, Cham (2015). DOI 10.1007/978-3-319-14791-8
17. Engelbrecht, J., Peets, T., Tamm, K., Laasmaa, M., Vendelin, M.: On the complexity of signal propagation in nerve fibres. Proc. Estonian Acad. Sci. **67**(1), 28–38 (2018). DOI 10.3176/proc.2017.4.28
18. Engelbrecht, J., Salupere, A., Tamm, K.: Waves in microstructured solids and the Boussinesq paradigm. Wave Motion **48**(8), 717–726 (2011). DOI 10.1016/j.wavemoti.2011.04.001

19. Engelbrecht, J., Tamm, K., Peets, T.: On mathematical modelling of solitary pulses in cylindri-
    cal biomembranes. Biomech. Model. Mechanobiol. **14**, 159–167 (2015). DOI 10.1007/s10237-
    014-0596-2
20. Engelbrecht, J., Tamm, K., Peets, T.: On solutions of a Boussinesq-type equation with
    amplitude-dependent nonlinearities: the case of biomembranes. arXiv:1606.07678 [nlin.PS]
    (2016)
21. Engelbrecht, J., Tamm, K., Peets, T.: On solutions of a Boussinesq-type equation with
    displacement-dependent nonlinearities: the case of biomembranes. Philos. Mag. **97**(12), 967–
    987 (2017). DOI 10.1080/14786435.2017.1283070
22. Engelbrecht, J., Tobias, T.: On a model stationary nonlinear wave in an active medium. Proc.
    R. Soc. London **A411**, 139–154 (1987)
23. FitzHugh, R.: Impulses and physiological states in theoretical models of nerve membrane.
    Biophys. J. **1**(6), 445–466 (1961). DOI 10.1016/S0006-3495(61)86902-6
24. Fornberg, B.: A Practical Guide to Pseudospectral Methods. Cambridge University Press,
    Cambridge (1998)
25. Fornberg, B., Sloan, D.: A review of pseudospectral methods for solving partial differential
    equations. Acta Numer. **3**, 203–267 (1994). DOI 10.1017/S0962492900002440.
26. Heimburg, T., Jackson, A.D.: On soliton propagation in biomembranes and nerves. Proc. Natl.
    Acad. Sci. USA **102**(28), 9790–9795 (2005). DOI 10.1073/pnas.0503823102
27. Hindmarsh, A.: ODEPACK, A Systematized Collection of ODE Solvers, vol. 1. North-Holland,
    Amsterdam (1983)
28. Hodgkin, A.L.: The Conduction of the Nervous Impulse. Liverpool University Press (1964)
29. Hodgkin, A.L., Huxley, A.F.: A quantitative description of membrane current and its appli-
    cation to conduction and excitation in nerve. J. Physiol. **117**(4), 500–544 (1952). DOI
    10.1113/jphysiol.1952.sp004764
30. Jones, E., Oliphant, T., Peterson, P.: SciPy: Open Source Scientific Tools for Python (2007).
    URL http://www.scipy.org
31. Lieberstein, H.: On the Hodgkin-Huxley partial differential equation. Math. Biosci. **1**(1),
    45–69 (1967). DOI 10.1016/0025-5564(67)90026-0.
32. Maugin, G.A.: Nonlinear Waves in Elastic Crystals. Oxford University Press, Oxford (1999)
33. Maugin, G.A.: Solitons in elastic solids (1938–2010). Mech. Res. Commun. **38**(5), 341–349
    (2011). DOI 10.1016/j.mechrescom.2011.04.009
34. Maurin, F., Spadoni, A.: Wave propagation in periodic buckled beams. Part I: Ana-
    lytical models and numerical simulations. Wave Motion **66**, 190–209 (2016). DOI
    10.1016/j.wavemoti.2016.05.008
35. Maurin, F., Spadoni, A.: Wave propagation in periodic buckled beams. Part II: Experiments.
    Wave Motion **66**, 210–219 (2016). DOI 10.1016/j.wavemoti.2016.05.009
36. Mindlin, R.: Micro-structure in linear elasticity. Arch. Ration. Mech. Anal. **16**(1), 51–78
    (1964).
37. Morris, C., Lecar, H.: Voltage oscillations in the barnacle giant muscle fiber. Biophys. J. **35**(1),
    193–213 (1981). DOI 10.1016/S0006-3495(81)84782-0
38. Nagumo, J., Arimoto, S., Yoshizawa, S.: An active pulse transmission line simulating nerve
    axon. Proc. IRE **50**(10), 2061–2070 (1962). DOI 10.1109/JRPROC.1962.288235
39. Neu, J.C., Preissig, R., Krassowska, W.: Initiation of propagation in a one-dimensional ex-
    citable medium. Phys. D Nonlinear Phenom. **102**(3-4), 285–299 (1997). DOI 10.1016/S0167-
    2789(96)00203-5
40. Peets, T., Tamm, K.: On mechanical aspects of nerve pulse propagation and the Boussinesq
    paradigm. Proc. Estonian Acad. Sci. **64**(3), 331 (2015). DOI 10.3176/proc.2015.3S.02
41. Peets, T., Tamm, K.: Mathematics of nerve signals. In: A. Berezovski, T. Soomere (eds.)
    Applied Wave Mathematics II, *Mathematics of Planet Earth*, vol. 6, pp. 207–238. Springer,
    Cham (2019). DOI 10.1007/978-3-030-29951-4_10
42. Peets, T., Tamm, K., Engelbrecht, J.: Numerical investigation of mechanical waves in biomem-
    branes. In: S. Elgeti, J.W. Simon (eds.) Conf. Proc. YIC GACM 2015 3rd ECCOMAS Young
    Investig. Conf. 6th GACM Colloquium, July 20-23, 2015, Aachen, Ger., pp. 1–4 (2015)

43. Peets, T., Tamm, K., Engelbrecht, J.: On the role of nonlinearities in the Boussinesq-type wave equations. Wave Motion **71**, 113–119 (2017). DOI 10.1016/j.wavemoti.2016.04.003

44. Peets, T., Tamm, K., Simson, P., Engelbrecht, J.: On solutions of a Boussinesq-type equation with displacement-dependent nonlinearity: A soliton doublet. Wave Motion **85**, 10–17 (2019). DOI 10.1016/j.wavemoti.2018.11.001

45. Perez-Camacho, M.I., Ruiz-Suarez, J.: Propagation of a thermo-mechanical perturbation on a lipid membrane. Soft Matter **13**, 6555–6561 (2017). DOI 10.1039/C7SM00978J

46. van der Pol, B.: LXXXVIII. On "relaxation-oscillations". London, Edinburgh, Dublin Philos. Mag. J. Sci. **2**(11), 978–992 (1926). DOI 10.1080/14786442608564127

47. Porubov, A.V.: Amplification of Nonlinear Strain Waves in Solids. World Scientific, Singapore (2003)

48. Randrüüt, M., Braun, M.: On identical traveling-wave solutions of the Kudryashov-Sinelshchikov and related equations. Int. J. Non. Linear. Mech. **58**, 206–211 (2014). DOI 10.1016/j.ijnonlinmec.2013.09.013

49. Reissig, R., Sansone, G., Conti, R.: Qualitative Theorie Nichtlinearer Differentialgleichungen. Edizioni Cremonese, Roma (1963)

50. Salupere, A.: The pseudospectral method and discrete spectral analysis. In: E. Quak, T. Soomere (eds.) Applied Wave Mathematics, pp. 301–334. Springer Berlin Heidelberg, Berlin (2009). DOI 10.1007/978-3-642-00585-5

51. Schwiening, C.J.: A brief historical perspective: Hodgkin and Huxley. J. Physiol. **590**(11), 2571–2575 (2012). DOI 10.1113/jphysiol.2012.230458

52. Scott, A.C.: Nonlinear Science. Emergence and Dynamics of Coherent Structures. Oxford University Press (1999)

53. Sterratt, D., Graham, B., Gillies, A., Willshaw, D.: Principles of Computational Modelling in Neuroscience. Cambridge University Press, Cambridge (2011). DOI 10.1017/CBO9780511975899

54. Strogatz, S.H.: Nonlinear Dynamics and Chaos : with Applications to Physics, Biology, Chemistry, and Engineering. CRC Press (1994)

55. Tamm, K., Peets, T.: On solitary waves in case of amplitude-dependent nonlinearity. Chaos, Solitons & Fractals **73**, 108–114 (2015). DOI 10.1016/j.chaos.2015.01.013.

56. Taniuti, T., Nishihara, K.: Nonlinear Waves. Pitman, Boston (1983)

57. Thompson, J.: Instabilities and Catastrophes in Science and Engineering. Wiley, New York (1982)

58. Vargas, E.V., Ludu, A., Hustert, R., Gumrich, P., Jackson, A.D., Heimburg, T.: Periodic solutions and refractory periods in the soliton theory for nerves and the locust femoral nerve. Biophys. Chem. **153**(2-3), 159–67 (2011). DOI 10.1016/j.bpc.2010.11.001.

59. Wolfram Research: Mathematica, Version 11.3. Champaign, IL (2018)

Chapter 7
Physical Mechanisms

> *The physicist, in his study of natural phenomena, has two methods of making progress: (1) the method of experiment and observation, and (2) the method of mathematical reasoning. ... the latter enables one to infer results about experiments that have not been performed.*
>
> *Paul Dirac, 1939*

7.1 Basic Elements of an Ensemble of Waves

The whole ensemble of waves is the following: (i) action potential AP and the corresponding voltage Z associated with the ion current(s), here denoted by J; (ii) longitudinal wave LW in the biomembrane with the amplitude U ; (iii) pressure wave PW in the axoplasm with the amplitude P; (iv) transverse displacement TW with the amplitude W; (v) temperature change Θ. Note that above just one ion current is listed as a variable. The crucial question is: what are the physical mechanisms which link these signal components into a whole?

The starting point of the analysis is related to causality. The classical understanding in axon physiology is the HH paradigm: the whole process is electricity-centred and starts with generating the AP. However, according to the original HH model, the accompanying effects (see Chap. 5) are not considered. Later in Chap. 8 we follow the HH paradigm as the fundamental approach in contemporary axon physiology [5, 6] but try to link it to the accompanying effects. This means that one should pay attention to electrical-to-mechanical and mechanical-to-electrical couplings and to possible heat production.

7.2 Qualitative Observations from Experiments

We start here listing the experimental phenomena and discuss the effects and the proposed mechanisms which could cause these effects [13]. This analysis permits to list the reasons for possible interactions between the components of a signal in axons.

The following observations can be noted from the published experiments:
- The measured TW is bi-polar for the squid giant axon and its peak coincides with the peak of the AP [39], both have approximately the same duration.
- The measured TW is close to uni-polar for the garfish olfactory nerve and its peak

© The Author(s), under exclusive license to Springer Nature Switzerland AG 2021
J. Engelbrecht et al., *Modelling of Complex Signals in Nerves*,
https://doi.org/10.1007/978-3-030-75039-8_7

coincides with the peak of the AP [41].
- The peak of the force developed at the axon surface coincides fairly accurately with the peak of the AP [39].
- The peak of the PW lags behind the peak of the AP for the squid giant axon [42].
- Mechanical and thermal signals are in phase with the voltage changes [16].
- The shape and the width of a TW are similar to those of the measured AP (without the overshoot) for the rat neuron [43].
- The experiments with *Chara braunii* cells have demonstrated that the mechanical pulse (out-of-plane displacement of the cell surface) propagates with the same velocity as the electrical pulse and is (in most cases) of bi-polar nature [15].
- The AP and temperature Θ for the garfish olfactory nerve are almost in phase and the duration of the positive phase of heat production is very close to the duration of the depolarising phase of the AP [39].
- The AP is narrower than the temperature change and the thermal response cannot be directly proportional to the change of voltage [36].
- Good correlation exists between the initial positive heat and the potassium (K^+) leakage [21].
- The residual heat exists after the passage of the AP and it is absorbed in time [35].

Possible mechanisms of coupling

The physical mechanism governing the HH model is based on the flow of ions through the biomembrane upon the change of ion concentration in the axoplasm. Contemporary understanding is that the ion channels may be voltage-gated like in the HH model but also mechanically-sensitive [30, 34]. This must be taken into account in building up a fully coupled model. Note that ion channels can also be sensitive to ion concentrations but this idea needs a mathematical formulation before including it into the model.

Gross et al. [17] have analysed electromechanical transductions in nerves. For electrical-to-mechanical transductions, the mechanisms of electrostriction and piezo-electricity are analysed and argued that both mechanisms could predict the swelling effects. For mechanical-to-electrical transductions, it is proposed that the stress-induced changes due to surface charges influence the intracellular electric field.

A promising mechanism for coupling the electrical and mechanical signals is the flexoelectric effect which is manifested in the deformation of the biomembrane curvature under an imposed electric field [33]. The flexoelectric effect is used by Chen et al. [4] for modelling the coupling of the AP and mechanical wave (TW). The classical HH model combined with cable theory includes density change in the biomembrane induced by the flexoelectricity. The biomembrane is modelled as an elastic (or viscoelastic) tube where the flexoelectric force is included into the governing equation. This force depends on the local change in the membrane potential. The changes in the axon diameter are taken into account and the system is reciprocal – it can be triggered either by an electrical pulse resulting in an AP or by a mechanical stimulus.

A coupled model based on the primary AP which generates all other effects is proposed by El Hady and Machta [8]. A Gaussian profile (a pulse) for an AP is taken as a basic without modelling. The assumption is made that in the fibre the membrane has potential energy, and axoplasmic fluid has kinetic energy. The idea is that a surface wave (meaning the surface of the fibre) is generated in the membrane and in the bulk field within the axon the linearised Navier-Stokes equations are used for calculating the pressure. The ensemble includes voltage (pulse, not a typical AP), radial membrane displacement TW and the lateral displacement inside the axon (i.e., the PW). The heat is assumed to be produced as additional release of mechanical energy, summing transverse changes in the diameter and lateral stretch. So the sequence: AP – mechanical waves – heat (temperature) is followed.

Rvachev [37] has proposed that the axoplasmic pressure pulse (PW) triggers all the process. The PW triggers the Na^+ channels and the local HH voltage spike develops which in its turn opens the Ca^{2+} channels. Free intracellular Ca^{2+} activates then the contraction of filaments in the axoplasm which gives rise to the radial contraction in the lipid bilayer. The similar idea that a PW could cause the excitement is proposed by Barz et al. [2].

Based on experimental results, Terakawa [42] has suggested that the pressure PW arises either from a change in electrostriction across the axoplasm or from a change in charge-dependent tension along the axoplasm. He states that pressure response is correlated with membrane potential and not with the membrane current. A slight influence of electro-osmotic water flow to pressure response is detected.

Abbott et al. [1] discuss the heat production in Maia nerves. They give three possible reasons for heat production: (i) the positive heat is derived from the energy released during the rising phase of the AP and the negative heat due to the absorption of energy during the falling phase of the AP; (ii) the positive heat is due to the interchange of Na^+ and K^+ ions and the negative heat represents the partial reversal of this interchange; (iii) the heat production is related to exothermic and endothermic chemical reactions. Richie and Keynes [36] have supported similar experimental data like Abbott et al. [1]. They also stated that the energy of the membrane capacitor is proportional to the voltage square and that the thermal response cannot be directly proportional to the change of voltage.

Tasaki and Byrne [40] have analysed the heat production in bullfrog myelinated nerve fibres. They estimate theoretically the relation of voltage V to temperature Θ in dependence of time constant RC (capacity x resistance in an operational amplifier). For RC longer, they use V related to Θ and for RC shorter V – related to $d\Theta/dt$.

Here one must also mention the studies on acoustic pulses in lipid monolayers [32] and bilayers [31]. It is demonstrated that electrical and chemical changes (PH aspects) are an inseparable part of acoustic pulses in lipid interfaces. The constitutive equation that governs the propagation of an acoustic pulse (mechanical wave) models the electro-mechano-thermo-chemical coupling at the interface. The acoustic pulse has a similarity to the action potential like qualitative pulse shape, or an 'all-or-none' behaviour as well as annihilation upon collision [32]. The acoustic phenomena at phospholipid monolayers can be described also by a nonlinear fractional wave equation [23] which describes how a wave in the monolayer is influenced by the viscous coupling of an interface to the environment. This might be an essential step forward to improve the modelling of the full wave ensemble because it will account for the influence of the intracellular fluid. Clearly, the studies on acoustic pulses in bilayers [23, 31, 32] cast more light to the understanding of coupling effects in nerve fibres.

Summary

To sum up, there is no consensus about the coupling and transduction of energy. Some studies concentrate upon the electromechanical transductions, some – upon the coupling of an electrical signal and temperature. This way or another, the qualitative experimental observations serve as guides in modelling. According to the HH paradigm, the process is triggered by an electrical stimulus which generates the AP but it is also proposed that the stimulus could be of a mechanical character. This could be the LW [19] or the PW [37]. The model by Chen et al. [4] based on using the flexoelectric effect involves the reciprocity of electrical and mechanical components of the process. The generation of the temperature during the process is associated either with the electrical signal [1, 40] or to mechanical effects [8]. However, the role of chemical reactions in producing temperature changes is also under discussion [1]. The theoretical analysis by Nogueira and Conde Garcia [7] has proposed that heat production at axoplasmic level is related to the Joule effect during the upstroke of an AP.

This way or another, in most of the studies (both experimental and theoretical), the coupling is associated to local changes of fields which cause changes in the whole system.

7.3 Modelling of Coupled Signals and Coupling Forces

The crucial issue is how to model the coupling. It is proposed that the main hypothesis for constructing the coupling forces could be [11]: all mechanical waves in axoplasm and surrounding biomembrane together with the heat production are generated due to changes in electrical signals (AP or ion currents) that dictate the functional shape of coupling forces. The seconding hypotheses are: the changes in the pressure wave may also influence the waves in biomembrane and mechanical waves may influence the AP and ion currents. This means reciprocity between the signal components. In many studies the changes are mentioned [3, 17, 40, 42] as reasons for interactions. And back to the history: the German physiologist Emil Du Bois-Reymond has noticed in the 19th century that "the variation of current density, and not the absolute value of the current density at any given time, acts as a stimulus to a muscle or motor nerve" [18]. This statement is called the Du Bois-Reymond law.

What is change? In mathematical terms changes mean either space (X) or time (T) derivatives of variables. This gives the glue for proposing the functional shapes of forces which at the first approximation could be described in the form of first-order polynomials of gradients or time derivatives of variables (Z_X, J_X, U_X, P_X, and Z_T, J_T, U_T, P_T). Such an approach involves also certain flexibility in choosing the model.

Further on, a possible approach in modelling is envisaged based on ideas described above [10, 11]. The processes which compose the leading effects in signal propagation can each be described by single model equations. In the coupled model these single equations are united into a system by coupling forces. For a proof of concept, the coupling could be modelled by a simpler approach: from the AP and ion currents to all the other effects.

It has been proposed that the process could be divided into primary and secondary components [12]. This is important from the viewpoint of the physical basis of processes in nerve fibres.

(a) The **primary** components are characterised by corresponding velocities and their mathematical models are derived from wave equation(s). These components are the AP, LW and PW.

(b) The **secondary** components are either derived from the primary components like the TW or their models are derived from basic equations which do not possess velocities like the temperature Θ. In this case, the diffusion-type equation could serve as a basic mathematical model.

On the basis of experimental studies (Sect. 7.2) it seems plausible that there are several physical mechanisms of coupling. This concerns electrical-mechanical (AP to PW and AP to LW) and electrical-thermal (AP to Θ) transduction. In essence, the mechanisms should also include feedback and coupling between all the components of the ensemble. However, these effects could be of more importance in pathological situations (axon dysfunction) or for a detailed understanding of neural communication and neural activity in general.

As stated above, the coupling forces could be described by changes in variables. Note that gradients (space derivatives) act along the axon and time derivatives across the membrane. Based on thermodynamics (see [36, 38]), for temperature changes one should consider also the possible effects of Z or Z^2 (alternatively J or J^2). In general terms the basic mechanisms starting from AP and ion current (variables Z and J) might be the following:

(i) electric-biomembrane interaction resulting in a mechanical response (LW, variable U);

(ii) electric-fluid (axoplasm) interaction resulting in a mechanical response (PW, variable P);

(iii) electric-fluid (axoplasm) interaction resulting in a thermal response (Θ).

7.4 Possible Interactions

However, as said above, the feedback from mechanical waves to other components of the whole ensemble might also influence the process. In more detail these interactions can be characterised [13]:

(a) influence from the AP:

    (1) Pressure change in axoplasm – Z_X, i.e., AP gradient along the axon axis can influence the pressure as a result of charged particles present in axoplasm reacting to the potential gradient along the axon;

    (2) Pressure change in axoplasm – Z_T, i.e., potential changes across the biomembrane can lead to a pressure change proportional to the potential change through electrically motivated membrane tension changes (also taken into account for LW, as this is accounted for here the influence from mechanical displacement is taken proportional to Z_T);

    (3) Pressure change in axoplasm – J_T , i.e., the axoplasm volume change from ion currents in and out of axon through the biomembrane plus the effect of possible osmosis which is assumed to be proportional to the ionic flows;

    (4) Density change in biomembrane – Z_T, i.e., electrically induced membrane tension change or flexoelectricity;

    (5) Density change in biomembrane – J_T, i.e., membrane deformation as a result of ionic flow through the membrane (ion channels deforming surrounding biomembrane when opening/closing); note that ion channels are not modelled here in the FHN model explicitly;

    (6) Temperature change – Z^2 or J^2, i.e., temperature increase from current flowing through the environment (power); this effect is related to the Joule heating;

    (7) Temperature change – J, i.e., the endothermic term is dependent on the integral of J which is taken to be proportional to the concentration of reactants which decays exponentially in time after the signal passage; needs a new kinetic equation to be added;

(b) influence from the pressure P:

    (1) Density change in biomembrane – P_T, i.e., membrane deformation (displacement) from the local pressure changes inside the axon;

    (2) Temperature change – P_T, i.e., reversible local temperature change: when pressure increases then temperature increases proportionally and when pressure decreases then temperature decreases proportionally, this happens in the same timescale as pressure changes;

    (3) Temperature – P, i.e., the irreversible local temperature increase from energy consumed by viscosity (friction) – this is actually time integral of P_T;

(c) influence from the mechanical wave U in biomembrane:

    (1) Action potential change – U_T, i.e., ion currents are suppressed when membrane density is increased and amplified when membrane density is decreased;

    (2) Temperature change – U_T, i.e., the reversible local temperature change, when density increases the local temperature is increased proportionally and when density decreases the local temperature decreases proportionally, this happens at the same timescale as the density changes in biomembrane;

(3) Temperature change – U, i.e., the irreversible local temperature increase from energy consumed by the added friction/viscosity term in longitudinal density change model – this is actually time integral of U_T which is proportional to U.

These are the plausible processes which can influence the dynamics of nerve pulse propagation. However, not all of the listed effects might be energetically at the same scale as far as the dynamics of the nerve pulses is concerned. At this stage, the possible influence of temperature changes might have on properties of fibres are disregarded as experimental observations are showing temperature changes low enough to be negligible for these processes. In addition, the possible changes in the fibre diameter which could change the physical properties are also disregarded being of many scales lower compared with leading effects. In general, the significance of the listed effects must be determined by experimental and theoretical studies in the future.

Abbott et al. [1] have discussed the production of heat during the propagation of an AP. They propose that "the positive heat is due to exothermic chemical reactions associated with the permeability cycle which accompanies the action potential, and the negative heat to endothermic chemical reactions involved in an early anaerobic stage of recovery". Tamm et al. [38] have proposed the usage of internal variables for describing these effects. This idea is explained below in more details.

Further, in Sect. 8.1.2, the structure of coupling forces based on descriptions above, will be specified.

7.5 The Concept of Internal Variables

Experimental studies of signals in nerve fibres as described above, deal with measurable phenomena emphasised at the macro level of fibres. As mentioned above, Abbott et al. [1] suggest that the heat changes can be associated to chemical reactions. In order to model such phenomena which are hidden from direct measurements, the concept of internal variables has been introduced in continuum mechanics. This idea is traced back to P. Duhem, P. Bridgman and J. Kestin (see the overviews by Maugin [25, 26]). Briefly, leaving aside the thermodynamical considerations, the concept means a clear distinction between variables that can be measured, i.e., that are observable, and variables that are hidden, called internal (see the brief description in [9, 26]).

Observable variables are measured field quantities like in the case of nerve signals the amplitude of the AP, amplitudes of the mechanical waves and the temperature change. The models for these variables are derived from balance laws including the energy balance for diffusive behaviour of temperature. Internal variables describe the influence of the underlying hidden processes and, in this way, compensate for the lack of the precise description [28, 29].

Two fundamental questions must be answered [9]: (i) What is the physical description of an internal variable? (ii) How to determine the governing time-dependent law for an internal variable (or internal variables)? As stated by Maugin [24], the physical description of an internal variable is often "a matter of decision at the outset from the part of the scientist". Such a decision depends mostly on time and space scales of underlying processes. In physiology, the phenomenological variables n, m, h introduced by Hodgkin and Huxley [20] are typical internal variables related to ion currents [27]. The formalism for deriving the governing equation for an internal variable is derived by Maugin [24, 25]. As a result, typically a kinetic equation governing internal variable α is derived (for details, see [25]). This is also a case for the HH model [20], which, although ingeniously proposed, is not derived from basic considerations. The possible derivation of governing equations which govern n, m, h using the ideas of the formalism mentioned above, is described by Maugin and Engelbrecht [27].

The limit values of these variables lie between zero and unity, i.e., between two levels. Each of them α_i ($\alpha_1 = n$, $\alpha_2 = m$, $\alpha_3 = h$) is described by an evolution equation [20]:

$$\dot{\alpha}_i = p_{i1}(1 - \alpha_i) - p_{i2}\alpha_i, \tag{7.1}$$

or equivalently by

$$\dot{\alpha}_i = -\frac{\alpha_i - \alpha_{i0}}{\tau_i}, \tag{7.2}$$

where

$$\alpha_{i0} = \frac{p_{i1}}{p_{i1} + p_{i2}}, \quad \tau_i = \frac{1}{p_{i1} + p_{i2}}. \tag{7.3}$$

Here, as above, the dot denotes the derivative with respect to time, α_{i0} is the equilibrium value and τ_i is the relaxation time, while p_{i1} and p_{i2} are the coefficients specified ingeniously by Hodgkin and Huxley [20] from experiments.

The possible structure of evolution equations demonstrates clearly the relaxation-type dynamics of internal variables. Note that the concept of internal variables is also used for describing the cardiac muscle contraction [14]. In this case, following the Huxley model [22], the active stress in myocardium is influenced by hierarchical internal variables. The hierarchy follows the sequence $Ca^{2+} \rightarrow$ activation parameter $A \rightarrow$ activation parameters of cross-bridges $\rightarrow$ active stress.

What next?

We have now described dynamical effects in nerves (Chap. 5), the mathematics used for describing the single effects (Chap. 6) on possible physical mechanisms (Chap. 7). It is a challenge to put the pieces together and build up a coupled mathematical model of all these effects into one system.

References

1. Abbott, B.C., Hill, A.V., Howarth, J.V.: The positive and negative heat production associated with a nerve impulse. Proc. R. Soc. B Biol. Sci. **148**(931), 149–187 (1958). DOI 10.1098/rspb.1958.0012

2. Barz, H., Schreiber, A., Barz, U.: Impulses and pressure waves cause excitement and conduction in the nervous system. Med. Hypotheses **81**(5), 768–72 (2013). DOI 10.1016/j.mehy.2013.07.049

3. Bishop, G.H.: Natural history of the nerve impulse. Physiol. Rev. **36**(3), 376–399 (1956). DOI 10.1152/physrev.1956.36.3.376

4. Chen, H., Garcia-Gonzalez, D., Jérusalem, A.: Computational model of the mechanoelectrophysiological coupling in axons with application to neuromodulation. Phys. Rev. E **99**(3), 032406 (2019). DOI 10.1103/PhysRevE.99.032406

5. Clay, J.R.: Axonal excitability revisited. Prog. Biophys. Mol. Biol. **88**(1), 59–90 (2005). DOI 10.1016/j.pbiomolbio.2003.12.004

6. Courtemanche, M., Ramirez, R.J., Nattel, S.: Ionic mechanisms underlying human atrial action potential properties: insights from a mathematical model. Am. J. Physiol. **275**(1), H301–H321 (1998)

7. de A. Nogueira, R., Conde Garcia, E.: A theoretical study on heat production in squid giant axon. J. Theor. Biol. **104**(1), 43–52 (1983). DOI 10.1016/0022-5193(83)90400-9

8. El Hady, A., Machta, B.B.: Mechanical surface waves accompany action potential propagation. Nat. Commun. **6**, 6697 (2015). DOI 10.1038/ncomms7697

9. Engelbrecht, J.: Questions About Elastic Waves. Springer International Publishing, Cham (2015). DOI 10.1007/978-3-319-14791-8

10. Engelbrecht, J., Peets, T., Tamm, K.: Electromechanical coupling of waves in nerve fibres. Biomech. Model. Mechanobiol. **17**(6), 1771–1783 (2018). DOI 10.1007/s10237-018-1055-2

11. Engelbrecht, J., Tamm, K., Peets, T.: Modeling of complex signals in nerve fibers. Med. Hypotheses **120**, 90–95 (2018). DOI 10.1016/j.mehy.2018.08.021

12. Engelbrecht, J., Tamm, K., Peets, T.: Primary and secondary components of nerve signals. arXiv:1812.05335 [physics.bio-ph] (2018)

13. Engelbrecht, J., Tamm, K., Peets, T.: On mechanisms of electromechanophysiological interactions between the components of signals in axons. Proc. Estonian Acad. Sci. **69**(2), 81–96 (2020). DOI 10.3176/proc.2020.2.03

14. Engelbrecht, J., Vendelin, M., Maugin, G.A.: Hierarchical internal variables reflecting microstructural properties: application to cardiac muscle contraction. J. Non-Equilibrium Thermodyn. **25**(2), 119–130 (2000). DOI 10.1515/JNETDY.2000.008

15. Fillafer, C., Mussel, M., Muchowski, J., Schneider, M.F.: Cell surface deformation during an action potential. Biophys. J. **114**(2), 410–418 (2018). DOI 10.1016/j.bpj.2017.11.3776

16. Gonzalez-Perez, A., Mosgaard, L., Budvytyte, R., Villagran-Vargas, E., Jackson, A., Heimburg, T.: Solitary electromechanical pulses in lobster neurons. Biophys. Chem. **216**, 51–59 (2016). DOI 10.1016/j.bpc.2016.06.005

17. Gross, D., Williams, W.S., Connor, J.A.: Theory of electromechanical effects in nerve. Cell. Mol. Neurobiol. **3**(2), 89–111 (1983). DOI 10.1007/BF00735275

18. Hall, C.W.: Laws and Models: Science, Engineering, and Technology. CRC Press, Boca Raton (1999)

19. Heimburg, T., Jackson, A.D.: On soliton propagation in biomembranes and nerves. Proc. Natl. Acad. Sci. USA **102**(28), 9790–5 (2005). DOI 10.1073/pnas.0503823102

20. Hodgkin, A.L., Huxley, A.F.: A quantitative description of membrane current and its application to conduction and excitation in nerve. J. Physiol. **117**(4), 500–544 (1952). DOI 10.1113/jphysiol.1952.sp004764

21. Howarth, J.V., Keynes, R.D., Ritchie, J.M.: The origin of the initial heat associated with a single impulse in mammalian non-myelinated nerve fibres. J. Physiol. **194**(3), 745–93 (1968). DOI 10.1113/jphysiol.1968.sp008434

22. Huxley, A.: Muscle structure and theories of contraction. Prog. Biophys. Biophys. Chem. **7**, 255–318 (1957). DOI 10.1016/S0096-4174(18)30128-8
23. Kappler, J., Shrivastava, S., Schneider, M.F., Netz, R.R.: Nonlinear fractional waves at elastic interfaces. Phys. Rev. Fluids **2**, 114804 (2017). DOI 10.1103/PhysRevFluids.2.114804
24. Maugin, G.A.: Internal variables and dissipative structures. J. Non-Equilibrium Thermodyn. **15**(2) (1990). DOI 10.1515/jnet.1990.15.2.173
25. Maugin, G.A.: Nonlinear Waves in Elastic Crystals. Oxford University Press, Oxford (1999)
26. Maugin, G.A.: The saga of internal variables of state in continuum thermo-mechanics (1893-2013). Mech. Res. Commun. **69**, 79–86 (2015). DOI 10.1016/j.mechrescom.2015.06.009
27. Maugin, G.A., Engelbrecht, J.: A thermodynamical viewpoint on nerve pulse dynamics. J. Non-Equilibrium Thermodyn. **19**(1) (1994). DOI 10.1515/jnet.1994.19.1.9
28. Maugin, G.A., Muschik, W.: Thermodynamics with internal variables part i. general concepts. J. Non-Equilibrium Thermodyn. **19**(3), 217–249 (1994). DOI 10.1515/jnet.1994.19.3.217
29. Maugin, G.A., Muschik, W.: Thermodynamics with internal variables. Part II. Applications. J. Non-Equilibrium Thermodyn. **19**(3) (1994). DOI 10.1515/jnet.1994.19.3.250
30. Mueller, J.K., Tyler, W.J.: A quantitative overview of biophysical forces impinging on neural function. Phys. Biol. **11**(5), 051001 (2014). DOI 10.1088/1478-3975/11/5/051001
31. Mussel, M., Schneider, M.F.: It sounds like an action potential: unification of electrical, chemical and mechanical aspects of acoustic pulses in lipids. J. R. Soc. Interface **16**, 20180743 (2019). DOI 10.1098/rsif.2018.0743
32. Mussel, M., Schneider, M.F.: Similarities between action potentials and acoustic pulses in a van der Waals fluid. Sci. Rep. **9**, 2467 (2019). DOI 10.1038/s41598-019-38826-x
33. Petrov, A.G.: Electricity and mechanics of biomembrane systems: Flexoelectricity in living membranes. Anal. Chim. Acta **568**(1-2), 70–83 (2006). DOI 10.1016/j.aca.2006.01.108
34. Ranade, S.S., Syeda, R., Patapoutian, A.: Mechanically activated ion channels. Neuron **87**(6), 1162–1179 (2015). DOI 10.1016/j.neuron.2015.08.032
35. Richie, J.: Energetic aspects of nerve conduction: The relationships between heat production, electrical activity and metabolism. Prog. Biophys. Mol. Biol. **26**, 147–187 (1973). DOI 10.1016/0079-6107(73)90019-9
36. Ritchie, J.M., Keynes, R.D.: The production and absorption of heat associated with electrical activity in nerve and electric organ. Q. Rev. Biophys. **18**(04), 451 (1985). DOI 10.1017/S0033583500005382
37. Rvachev, M.M.: On axoplasmic pressure waves and their possible role in nerve impulse propagation. Biophys. Rev. Lett. **5**(2), 73–88 (2010). DOI 10.1142/S1793048010001147
38. Tamm, K., Engelbrecht, J., Peets, T.: Temperature changes accompanying signal propagation in axons. J. Non-Equilibrium Thermodyn. **44**(3), 277–284 (2019). DOI 10.1515/jnet-2019-0012
39. Tasaki, I.: A macromolecular approach to excitation phenomena: mechanical and thermal changes in nerve during excitation. Physiol. Chem. Phys. Med. NMR **20**, 251–268 (1988)
40. Tasaki, I., Byrne, P.M.: Heat production associated with a propagated impulse in bullfrog myelinated nerve fibers. Jpn. J. Physiol. **42**(5), 805–813 (1992). DOI 10.2170/jjphysiol.42.805
41. Tasaki, I., Kusano, K., Byrne, P.M.: Rapid mechanical and thermal changes in the garfish olfactory nerve associated with a propagated impulse. Biophys. J. **55**(6), 1033–1040 (1989)
42. Terakawa, S.: Potential-dependent variations of the intracellular pressure in the intracellularly perfused squid giant axon. J. Physiol. **369**(1), 229–248 (1985). DOI 10.1113/jphysiol.1985.sp015898
43. Yang, Y., Liu, X.W., Wang, H., Yu, H., Guan, Y., Wang, S., Tao, N.: Imaging action potential in single mammalian neurons by tracking the accompanying sub-nanometer mechanical motion. ACS Nano **12**(5), 4186–4193 (2018). DOI 10.1021/acsnano.8b00867

Chapter 8
An Ensemble of Waves

By a model is meant a mathematical construct which, with the addition of certain verbal interpretations, describes observed phenomena. The justification of such a mathematical construct is solely and precisely that is expected to work – that is correctly to describe phenomena from a reasonable wide area. Furthermore, it must satisfy certain aesthetic criteria – that is, in relation to how much it describes, it must be rather simple.

John von Neumann, 1955

8.1 The Model

We collect now the coupled mathematical model based on our earlier studies [17, 20, 59]. This model unites the AP, the LW, the TW, the PW and Θ into one system using the coupling forces. The dimensionless formulation is used. As noted earlier index X denotes the spatial partial derivative and T – the temporal partial derivative.

The governing equations are presented first, followed by the discussion of the parameters and the coupling forces. The AP is governed by the FHN model [50]:

$$Z_T = DZ_{XX} - J + Z\left(Z - C_1 - Z^2 + C_1 Z\right),$$
$$J_T = \epsilon_1\left(C_2 Z - J\right). \tag{8.1}$$

The pressure wave PW is governed by a modified wave equation:

$$P_{TT} = c_2^2 P_{XX} - \mu_2 P_T + F_2(Z, J). \tag{8.2}$$

The longitudinal wave LW in the biomembrane is governed by the improved HJ (iHJ) model [15, 28]:

$$U_{TT} = c_3^2 U_{XX} + NUU_{XX} + MU^2 U_{XX} + NU_X^2 + 2MUU_X^2$$
$$- H_1 U_{XXXX} + H_2 U_{XXTT} - \mu_3 U_T + F_3(Z, J, P), \tag{8.3}$$

and the transverse displacement TW is calculated from LW as $W \propto U_X$ (drawing inspiration from the theory of rods) [16, 56]:

$$W = KU_X, \tag{8.4}$$

where K is a coefficient. The temperature Θ is governed by the classical heat equation:

$$\Theta_T = \alpha\Theta_{XX} + F_4(Z, J, U, P). \tag{8.5}$$

J. Engelbrecht et al., *Modelling of Complex Signals in Nerves*, https://doi.org/10.1007/978-3-030-75039-8_8

Coupling forces following discussion in Chap. 7 are the following:

$$F_2 = \eta_1 Z_X + \eta_2 J_T + \eta_3 Z_T, \tag{8.6}$$

$$F_3 = \gamma_1 P_T + \gamma_2 J_T - \gamma_3 Z_T, \tag{8.7}$$

$$F_4 = \tau_{11} Z^2 + \tau_2 (P_T + \varphi_2(P)) + \tau_3 (U_T + \varphi_3(U)) - \tau_4 \Omega, \tag{8.8}$$

where

$$\Omega_T + \epsilon_4 \Omega = \zeta J, \tag{8.9}$$

or

$$\Omega_T = \varphi_4(J) - \frac{\Omega - \Omega_0}{\tau_\Omega}, \quad \Omega_0 = 0, \quad \tau_\Omega = \frac{1}{\epsilon_4}, \tag{8.10}$$

and

$$\varphi_2(P) = \lambda_2 \int P_T \, dT, \quad \varphi_3(U) = \lambda_3 \int U_T \, dT, \quad \varphi_4(J) = \zeta \int J \, dT. \tag{8.11}$$

The block diagram of the proposed model is presented in Fig. 8.1 where the arrows denote interactions through coupling forces and boxes represent model equations for the phenomena noted therein. The leftmost box represents an initial condition for the AP taken with an amplitude above the threshold value.

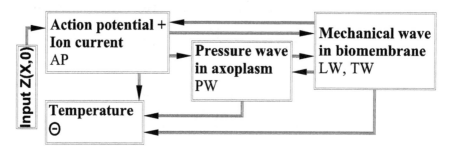

Fig. 8.1 The block diagram of the model. The arrows indicate the coupling which can include feedback.

Note that the whole signal in a nerve is an ensemble which includes primary and secondary components. The primary components are governed by wave-type equations characterised by corresponding velocities. These components are the AP, the longitudinal wave LW in the biomembrane and the pressure wave PW in the axoplasm. The secondary components have no specific velocities: the temperature Θ and the transverse displacement W of the biomembrane. The temperature change is governed by a diffusion-type equation while the transverse displacement is calculated from the longitudinal deformation of the biomembrane.

8.1.1 Notations, Variables and Parameters in the Model

Equations (8.1) are the FHN equations [50] which model the AP. The notations are the following: $C_i = a_i + b_i$ and $b_i = -\beta_i U$, Z is the voltage, J is ion current, ϵ_1 is the parameter governing the difference of time scales, a_i is the 'electrical' activation coefficient, b_i is the 'mechanical' activation coefficient (modelling the influence of the LW in the membrane to the ion channels) and U is a longitudinal density change from the lipid bi-layer density model (8.3).

Equation (8.2) is a modified wave equation with viscosity effect included, used for modelling the PW, where P is the pressure, μ_2 is the viscosity coefficient, F_2 is the coupling term accounting for the possible influence from the AP and TW.

Equation (8.3) governs the longitudinal wave in the lipid bilayer, modelled by the improved HJ model [15, 28], where $U = \Delta\rho$ is the longitudinal density change, c_3 is the sound velocity in the unperturbed state, N,M are nonlinear coefficients, H_i are dispersion coefficients and μ_3 is the dissipation coefficient. Note that H_1 accounts for the elastic properties of the bi-layer and H_2 – for the inertial properties. Term F_3 is the coupling force accounting for the possible influence from the AP and PW. The transverse displacement (TW) is $W \propto U_X$. In Eq. (8.3) a viscous term U_T is included as the lowest order dissipative term. This follows the idea by Kaufmann [40] who stressed the need to account for the dissipative processes in the biomembrane (see also [38, 44]). In soliton physics, the dissipative solitons are well known [7] because the real systems are not conservative like the classical soliton theory describes. Here including such a term means a step closer to reality. It should be noted that taking a higher-order dissipative term, like, for example, U_{XXT} as proposed by Lundström [44] is also a possibility (c.f. Eq. (3.41)).

Equation (8.5) governs the local temperature Θ modelled by the classical heat equation, where Θ is the temperature, α is the thermal conductivity coefficient and F_4 is the coupling force accounting for the possible influence from the AP, LW and PW. Note that in the present formulation the equilibrium temperature has been taken as zero level and Θ describes temperature change compared to the 'zero' level.

8.1.2 Coupling Forces

Our hypothesis on the coupling of effects is based on coupling forces [17]. This hypothesis unites mathematical and physical explanations. The physical considerations are analysed by Mueller and Tyler [49].

In coupling force F_2 (8.6), Z_X accounts for the presence of charged particles under the influence of the potential gradient (along the axon), J_T accounts for the ionic flows into and out of an axon (across the membrane) and Z_T accounts for the possible pressure change as a result of membrane tension changes from the electrical field and $\eta_i =$const.

In coupling force F_3 (8.7), P_T accounts for possible membrane deformation because of pressure changes (pressure to TW to LW), J_T accounts for possible membrane deformation as a result of ionic flows through ion channels and Z_T accounts for possible electrically induced membrane tension change [5, 54, 62] and γ_i =const. Note the sign, assuming that density decreases with the increasing tension.

The usage of the complicated formulation of coupling force F_4 is motivated by many possible mechanisms mentioned in experimental studies [1, 57]. Term Z^2 in (8.8) is related to $\mathcal{P} = Z^2/r$ where r is resistance and $\mathcal{P}$ is the power which accounts for the Joule heating from the (electrical) current flowing through the environment, Ω is an internal variable, modelling all endothermic processes and τ_i =const. In (8.8) the integrals for $\varphi_2(P)$ and $\varphi_3(U)$ in (8.11) characterise the thermal influence from mechanical waves into the temperature (8.5) as a result of friction–type terms in Eqs. (8.2) and (8.3) and λ_i are coefficients. In dimensionless case $\lambda_i = \mu_i$. In (8.8), term P_T accounts for the local temperature increase if the pressure increases and decrease when pressure decreases (reversible), term $\int P_T \, dT$ accounts for the temperature increase from the energy lost to viscosity in (8.2), term U_T accounts for the local temperature increase when the local density is increased and decrease if the local density decreases (reversible), term $\int U_T \, dT$ accounts for the temperature increase from the energy lost to viscosity in (8.3) and finally, as noted, the internal variable Ω takes into account the temperature change from endothermic processes. In (8.9), the term J is used on the assumption that endothermic process intensity is proportional to some kind concentration which is taken as the time integral of ion currents at the location (ζ is coefficient). This concentration-like quantity is decaying exponentially in time which is accounted by the term $-\epsilon_4\Omega$ in Eq. (8.9). In other words, we are considering some kind of an endothermic chemical reaction. In principle, it describes the change from one level to another like the phenomenological variables describing ion currents in the HH model [30]. In Eqs. (8.9), Ω_0 denotes the equilibrium level (in our setting equilibrium has been taken as zero level) and τ_Ω is the relaxation time. Clearly, Ω as an internal variable has a certain relaxation time before reaching its equilibrium value. Following the discussion of Abbott et al. [1], the thermal influence of chemical reactions might initially be endothermic and in the later stages of the recovery – exothermic. In principle, this single internal variable can be used to describe both endo– and exothermic chemical influence, which means that the value of Ω has to change the sign (overshooting equilibrium value). However, if these endo– and exothermal processes happen at different time scales it is clearer to use dual internal variables [19].

In the case of using dual internal variables, the selection of dependent variables must be carefully considered as there are several logical options. The mathematical model used for many simulations [17, 18] involves the FHN model for describing ion current J. In terms of the HH model, this ion current corresponds to the 'turning-on' and 'turning-off' within one variable. In the HH model, there are several ion currents and the profiles of two ion currents: sodium (J^{Na}) and potassium (J^K) (Fig. 6.1 right panel) were shown compared with the propagating AP (Fig. 6.1 left panel) [30]. Both ion currents have reciprocal signs which is a significant property. If now instead of an internal variable Ω in (8.8) two internal variables $\Omega^{ex}(J^{Na})$ and $\Omega^{en}(J^K)$ are introduced, then it is possible to specify the role of the exothermic and endothermic processes. In this case the coupling force F_4 (8.8) can be taken as

$$F_4 = \tau_1 Z^2 + \tau_2 \left(P_T + \varphi_2(P)\right) + \tau_3 \left(U_T + \varphi_3(U)\right) + \tau_{41}\Omega^{\text{ex}} - \tau_{42}\Omega^{\text{en}}, \qquad (8.12)$$

with

$$\Omega_T^{\text{ex}} + \epsilon_{41}\Omega^{\text{ex}} = \zeta_1 J^{\text{Na}}, \qquad \Omega_T^{\text{en}} + \epsilon_{42}\Omega^{\text{en}} = \zeta_2 J^{\text{K}}, \qquad (8.13)$$

and where τ_{41}, τ_{42}, ϵ_{41}, ϵ_{42}, ζ_1, ζ_2 are the coefficients. The J^{Na} and J^{K} are corresponding ion currents from the HH model. Note that in this case relaxation times $\tau_{\Omega^{\text{ex}}}$, $\tau_{\Omega^{\text{en}}}$ are different which corresponds to the temporal difference of exothermic and endothermic processes. Noting that in the HH model the ion currents J^{Na} and J^{K} are roughly simultaneous and roughly equal but with opposite signs (Fig. 6.1) it is possible to use the single ion current from the FHN model as an approximation to these currents. However, some details might be lost if the modelled process would be such that in the HH model these ion currents would be different in magnitude or with a significant difference in temporal scales.

8.2 Energetical Balance

The general understanding is that the whole process of signal propagation in nerve fibres is adiabatic and reversible [26]. However, the experiments have demonstrated that there is residual heat after the passage of an AP. This is demonstrated for mammalian (rabbit) nerves [32] and fish (pike) nerves [33], in both cases about 10% of positive initial heat. Heimburg [26] stated that "from the perspective of an energy, the heat change is in fact larger than the electrical effect, even if temperature changes are small".

The challenge is to understand the general energy E balance:

$$E = E_{\text{AP}} + E_{\text{PW}} + E_{\text{LW}} + E_\Theta, \qquad (8.14)$$

where indices denote the corresponding waves and temperature. Note that energy of ion currents [45] is hidden in the E_{AP} because actually, they play the role of internal variables in supporting the AP. The energy E_{AP} of an AP is [29, 57]

$$E_{\text{AP}} = \frac{1}{2} C Z^2, \qquad (8.15)$$

where C is the membrane capacitance and Z - the amplitude of the AP. In general, energy should be related to motion, i.e., kinetic energy (see also Barz et al. [2]). For wave equations, it means the dependence on the square of the amplitude A (i.e., on A^2) (see Margineanu and Schoffeniels [45] for the energy of ion currents). Heimburg and Jackson [27] have explicitly proposed that for the LW according to their model, the energy is

$$E_{\text{LW}} = \frac{c_0^2}{\rho_0^A} U^2 + \frac{p}{3}\rho_0^A U^3 + \frac{q}{6}\rho_0^A U^4, \qquad (8.16)$$

where c_0 is the velocity, ρ_0^A is the density, U is the amplitude and p, q are the nonlinear parameters (see also [49]). Note that nonlinearity affects also the energy. The more general analysis of energy in biomembranes [10] in terms of continuum mechanics has divided the surface Helmholtz energy to local and non-local components. The local components depend on deformation and the non-local ones – on curvatures of the biomembrane. El Hady and Machta [12] proposed that the energy of the biomembrane is related mostly to potential energy U_{LW+TW} and the energy of the axoplasm – to kinetic energy T_{PW}.

It is also known that the energy in an electromagnetic wave is proportional to the square of its peak electric field. The dissipation needed for describing the real processes [40] is included also to the modified wave equations governing the longitudinal waves in biomembranes and axoplasm. As said by Margineanu and Schoffeniels [45] for ionic currents in the HH model, energy dissipation is degraded into heat. The temperature effects if described by equation (8.5) are dissipative by nature. So, from the viewpoint of energy balance, the coupled model described above in Sect. 8.1 can redistribute energy between its components although we presently do not know the transduction of energy in mathematical terms. For example, in case of the LW, expression (8.16) describes the conservative situation for an LW only. The total balance for the mechanical waves in the biomembrane (the LW and the TW) during the propagation of the coupled signal could be described by

$$E_{LW+TW}^{total} = E_{LW} + E_{LW}^{coupl.} - E_{LW}^{dissip.}, \tag{8.17}$$

where $E_{LW}^{coupl.}$ denotes energy inflow (through coupling force F_3) and $E_{LW}^{dissip.}$ denotes the energy loss from dissipation and through possible coupling with other components (for example, if energy exchange between PW and LW is accommodated). The similar arguing concerns the PW and Θ. One must agree with Heimburg [26] that new experiments are needed because there are many unanswered questions.

8.3 Simplifications

The system of governing equations and coupling forces (8.1) to (8.8) is quite complicated involving many possible mechanisms which might affect nerve pulse propagation and associated effects. Each of these elements has some physical justifications for its presence but are they all really needed? This is an important question. For example, some of these might have a negligible enough effect on the evolution of the wave ensemble in reality and therefore they might be dropped from the governing equations. However, this is a question that can not be answered convincingly without some further experimental evidence. What can be done is outlining some simplifications for the system and circumstances under which such a simplification would make sense. In addition to simplifications to the governing equations, another avenue to be studied is the question whether all the effects included in the coupling forces are significant enough to be taken into account.

8.3.1 Action Potential

The FHN model (8.1) used is already a simplification of the HH model. There is not much room for simplifying the equation further without losing the essential characteristics of the AP. For example, using an evolution equation [13] for the AP would mean that one loses the pulse annihilation during the head-on collision because evolution equations do not support simultaneous counter-propagating solutions. However, under circumstances where the head-on collision is not considered, such a simplification might be rational.

Whatever the simplification, the key point is that the model framework only needs an action potential signal shape and some kind of estimate for the ionic flow. One could use, for example, even an experimentally measured AP profile and use the procedure outlined in the HH model to derive an ion current from that measured profile. Technically this could mean using the classical wave equation for the AP by giving the measured profile as an initial condition, although doing so means losing some relevant effects, like the above-mentioned pulse annihilation upon the head-on collision but also any feedback from other constituents of an ensemble that should affect the AP. It should be noted as well that the AP might not have a constant velocity and a shape throughout its entire life when propagating along the axon [9].

In the used FHN model one possibility of simplification is to drop the 'mechanical' activation coefficients b_i if the mechanical wave in the biomembrane is assumed to have a negligible influence on the ion channel functionality.

8.3.2 Pressure Wave in Axoplasm

The pressure wave in axoplasm is already modelled by the classical wave equation with added viscosity. Only the viscosity effect could be dropped if the process duration and propagation distances are assumed to be so short that the effect of viscosity can be neglected. If viscosity is dropped then the temperature coupling force (8.8) should be simplified as well by removing the temperature contribution from pressure wave to the temperature through friction-like forces. Where there might be more room for simplifications, is the coupling force (8.6). For example, Terakawa [62] has noted that the pressure response in the axoplasm is suppressed by some anaesthetics (like lidocaine) and has concluded that the pressure response is free from components of the axoplasmic origin. In terms of the proposed framework, this could mean, for example, focusing on the Z_T in (8.6) as the dominating source of the pressure response and either scaling back or even neglecting the other two contributions. However, the rationality of such a simplification should be established by further studies because more recent experimental papers investigating the pressure characteristics in axons with more modern techniques have not materialised.

8.3.3 Longitudinal Wave in Biomembrane

The improved Heimburg-Jackson model (8.3) is used in the present framework because it had a connection to experimental measurements when the model framework was initially proposed. Initially, the model framework consisted only the AP and mechanical wave in a biomembrane. However, in principle this model is a Boussinesq-type equation [6] with an unusual nonlinearity, meaning that like all Boussinesq-type equations it could be simplified to its basic core which is the classical wave equation if all the nonlinear and dispersive effects are neglected. The dissipation should be probably the last effect dropped as it seems to be the most essential to describe the experimental observations because of synchronisation of wave velocities in a propagating ensemble (see, for example, [21]). Without the dissipation the mechanical wave is free to travel at the sound velocity in a given environment, which is much higher in the environment with roughly the density of the water than the typical propagation velocities of the AP, meaning the observed wave speed synchronisation would need to have some other explanation.

For the coupling force (8.7), one of the possible simplifications could be neglecting the contribution of the P_T as Terakawa [62] has suggested that the mechanical response arises mainly from the phospholipid region of the membrane. He argues that the electrostriction and the Kerr effect are both expected to cause membrane thickness changes with the observed membrane potential dependence and another factor that could contribute to mechanical responses is a potential-dependent change in membrane tension. However, as noted earlier, such a simplification could use some further justifications based on experiments with more modern techniques and with a wider range of test subjects than only a giant squid.

8.3.4 Temperature

The classical heat equation is the simplest way to model temperature. It is possible, in principle, that the temperature changes caused by the propagating nerve pulse are so rapid compared to the normal temperature transfer through diffusive processes that the thermal conductivity coefficient could be set to zero making the temperature strictly local quantity if the processes under investigation are very fast. However, doing so should be justified by separate estimates as such a simplification is in principle nonphysical because all real environments conduct heat.

Most of the simplifications in regard to the temperature likely involve reassessing the need for all the noted thermal sources in the coupling force. The Joule heating seems to be the most essential because of the main hypothesis made that the whole process starts with an electrical signal [17, 59] although this would need to be assessed for the specific conditions under modelling. Second, mechanical effects could be small enough to be disregarded if experimental observations support that or a separate assessment finds them to be negligibly small. The internal variables introduced are needed if the temperature response needs to be happening at different timescale than the driving signal. Only the endothermic variant of the internal variable may be sufficient for explaining experimental observations, as it might be difficult to separate the exothermal internal variable from all the other sources of heat without making some additional assumptions about underlying mechanisms. The minimum set of effects for coupling force F_4 which can describe experimental observations (like, for example, [1]) could be to have only the Joule heating and endothermic internal variable.

8.3.5 Neglecting Effects

In addition to the possibility of simplifying the proposed governing equations, another possibility is neglecting some effects if under some circumstances these can be demonstrated to be negligible. The model framework is built up by a quite straight-forward process and to neglect one or another effect (and its governing equation) is possible. Such simplifications, however, should be separately justified as all the effects having governing equations are experimentally verified to exist. A minimal system of governing equations would be the FHN equation describing only the action potential. Because the AP is assumed to be the driving signal for all mechanical and thermal components in an ensemble dropping the AP would mean a reassessment of all assumptions done during the modelling which falls outside of the proposed framework, even if it would be technically possible from the mathematical standpoint. Actually, it is possible to take the mechanical signal in the lipid bilayer as the driving force because large enough deformation of the biomembrane can push the AP above the threshold and lead into emerging of the AP through the 'mechanical' activation coefficients b_i included in the FHN model. Note that El Hady and Machta [12] have used just a Gaussian pulse for the AP without any governing equation, as an example.

Two component simplifications

The next possible simplification is two-component models where only two aspects are taken into account. All combinations seem plausible if other effects can be shown to be negligible enough to be dropped [14]. The possible combinations are:

○ *The AP and the temperature* Θ

When the mechanical effects are considered negligible, the Joule heating from AP is enough to generate the heat signal and even an internal variable can be included as it is a function of the ion current from the AP model.

○ *The AP and the longitudinal wave (LW)*

The next logical possibility is considering only the AP and the density wave in a biomembrane. It is a logical simplification, as in experiments these two are often the only signals tracked. The experimentally observed temperature changes are small enough to not affect signal propagation in a meaningful way, at first glance, and could be rather considered as an additional source of information about the process than an effect with measurable feedback to the system evolution. The influence of the internal pressure changes is a more open question but one could assume, for example, these to be just proportional to the surface deformation (as the lipid bilayer is very thin and could be considered almost ideally elastic) so some justifications can be found to not track the internal pressure.

○ *The AP and the pressure wave (PW)*

Similarly to the AP and the lipid bi-layer simplifications, the arguments made in regards of possibly dropping the pressure can be used to argue in favour of dropping the waves in the lipid bilayer as a model component and argue that the membrane is thin and elastic. Then the tracking of only the internal pressure in axon could be sufficient while assuming the surface displacement to be directly proportional to the changes in the internal pressure.

○ *The longitudinal wave (LW) and the pressure wave (PW)*

When the focus is on an experimental setup which does not involve the AP it is possible to consider only the mechanical waves. For example, one could somehow change the internal pressure of the axon and track the resulting pressure wave and membrane displacement caused only by the internal pressure changes. This simplification does not seem plausible for normal neural activity modelling considering the overwhelming experimental evidence with regard to the importance of the AP in the context of nerve function.

○ *The mechanical wave (PW or LW) and the temperature* Θ

Similarly to the arguments made before it is possible to consider a case where only one of the mechanical waves is tracked (the pressure wave or the wave in the lipid bilayer) on assumption that the other mechanical wave is proportional to the one tracked and combine this with the tracking of temperature. While this might be not relevant for the normal neural activity it could allow experimental assessment of the importance or even relevance of the temperature contributions from the mechanical effects.

Three component simplifications

The same arguments made in favour of dropping two effects from the model framework can be used to justify only dropping a single effect from the tracked ensemble.

8.4 Modifications

The mathematical model presented in Sect. 8.1 is certainly a robust one. It includes the governing equations and coupling forces which model the energy transfer between the components of the wave ensemble. The structure of coupling forces is based on the hypotheses introduced in Chap. 5 and their possible effects described in Chap. 7. It is quite clear that all the equations and coupling forces may be modified to better match the reality. Further, we analyse briefly the possible modifications bearing in mind that above we have followed the HH paradigm. That means that possible developments of the soliton paradigm are not discussed.

8.4.1 Properties of the Environment

An axon is placed into an extracellular fluid that has a certain concentration of ions. Such an environment plays certainly a role for ion currents through the lipid biomembrane. The experimental studies [23, 25] have revealed the behaviour of lipid ion channels under various conditions but the question of how to use this information in mathematical models of propagation of an AP is still a challenge. Besides the ion concentration of the extracellular fluid, the temperature of the environment should be taken into account. This effect has been established already in earlier studies of nerve pulses [31, 61] and is depicted in Fig. 8.2.

Fig. 8.2 Records of action potentials at three temperatures, superimposed on the same baseline and stimulus artefact. Temperatures and amplitudes, respectively: (A) 32.5 °C, 74.5 mV; (B) 18.5 °C, 99 mV; (C) 5 °C, 108.5 mV. Time marks: 1 msec. Reproduced with permission from [31]; ©John Wiley and Sons 1949 (colours inverted for better visibility).

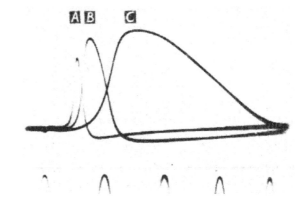

Tasaki and Fujita [61] have demonstrated that the temperature change of the environment leads to changes in (i) the conduction time; (ii) the spike height; (iii) the spike duration; (iv) the strength-latency relation. The straightforward way to take such a dependence into account is to use the temperature-dependent parameters of the governing equations. The numerical simulations have used the dimensionless forms to catch the qualitative profiles but if the physical units are used then the parameters could be taken as temperature-dependent. Scott [58] has analysed the HH model at a temperature of 6.3 °C. He proposed that at other temperatures the rates of changing n, m, h should be multiplied by factor κ, where

$$\kappa = 3^{(T-6.3)/10}. \tag{8.18}$$

The temperature-dependence can be introduced also in the FHN model [4] and the results are qualitatively similar to earlier experimental observations [31, 61] (see Fig. 8.2). Bini et al. [4] have demonstrated that at low temperatures the AP in the FHN model has a larger duration and on varying the temperature the amplitude of the AP becomes smaller. In this case, the experimental validation is extremely important.

Besides the temperature effects, the viscosity of the extracellular fluid can influence the mechanical waves in a biomembrane. The studies of phospholipid monolayers [39] have demonstrated that in this case, a suitable mathematical model describing the wave process in the system interface-bulk medium is a fractional wave equation. This equation models a so-called Lucassen surface wave [43] which takes the viscosity of the adjacent bulk medium into account. It might be interesting to combine this idea with the model of the biomembrane (Eq. (8.3)) as a step towards a better description of reality.

8.4.2 Properties of Structures

We have considered the biomembrane as a lipid bilayer and the axoplasm as a viscous fluid. However, there are numerous cellular and molecular players which influence the neural functions [49]. The list is long: stretch-sensitive ion channels which make a biomembrane inhomogeneous; cytoskeletal proteins which may influence the opening and closing the ion channels; extracellular matrix proteins; actin filaments in the axoplasm which probably influence the homogeneity of the axoplasm, etc. The capacity of the biomembrane may be changed during the process, etc. Certainly, there are more ions than in the HH model (K^+ and Na^+), for example, Ca^{2+} ions may play an important role [48]. In this case, the kinetic equations for phenomenological variables n, m, h in the HH model must be replaced by new variables M, N with different governing equations (see Sect. 6.1). It has been shown [36] that protein transport, lipid bi-layer phase, membrane pressure and stiffness can influence the propagation process. One should consider also general oxygen consumption, CO_2 output, the influence of carbohydrates which all could be of importance under certain conditions. One measured effect is the change of the radius of fibre (the TW) and there is a question whether to take this change into account in the model by a certain feedback mechanism which could change the parameters. The measurements have demonstrated that the amplitude of the TW (swelling) is at the nanometer scale [34, 60, 63] and can be neglected in comparison to the fibre diameter. However, if needed, then in principle such a feedback mechanism could be developed.

The present model deals with the unmyelinated biomembrane. An important improvement of the model could be the modification where the myelin sheath could be taken into account. It means the changes in the elastic and inertial properties of the biomembrane with considerable inhomogeneities (Ranvier nodes). The overviews by Meissner [47] and by Drukarch et al. [11]) reveal many electrophysiological aspects that should be taken into account in modelling in order to enhance the predictive power of models.

8.4.3 Mechanisms

In Chap. 7, the mechanisms of interactions were analysed in detail: (i) electric-lipid bi-layer interaction resulting in the mechanical wave in biomembrane; (ii) electric-fluid interaction resulting in the mechanical wave in the axoplasm; (iii) electric-fluid interaction resulting in the temperature change in axoplasm. These interactions were described by changes in fields realised by derivatives of variables. Although such mechanisms seem to grasp the essential features of coupling of signals, there are many possibilities more, especially in the case of pathological situations. Many studies are demonstrating the possibilities of using biomechanical effects for explaining the experimental results. The swelling of biomembranes could be explained based on the flexoelectric effect [54, 55]. This is done by Chen et al. [5] for linking the AP with the TW in the biomembrane. Their model includes: (i) the HH model for calculating the AP; (ii) a model for the mechanical wave in the axonal cylinder (either elastic or viscoelastic); (iii) a reverse flexoelectric model linking membrane bending to changes in the AP; (iv) a direct flexoelectric model linking the membrane polarisation to the membrane mechanical strain gradient. Such a model could also be used for explaining effects in spinal cord injury [37]. Gross et al. [24] have analysed the mechanism of mechanical-to-electrical and electrical-to-mechanical transductions. In the first case, it is proposed that the surface charge on the biomembrane causes stress-induced changes in the intramembrane electric field. The reason for the second transduction is attributed to either electrostriction or piezoelectricity. The last reason is suggested also by Terakawa [62] as a possibility for the transduction. The physical phenomenon called osmosis (electro-osmotic water flow through the membrane) is also studied within the context of nerve pulses. The influence of this effect has been estimated small [24, 62] and is therefore neglected. Like in the mathematical model proposed and advocated within this book, the transduction according to Chen et al. [5] or Gross et al. [24] is realised by coupling forces derived from the physical considerations.

8.4.4 Systems

The proposed framework has been focused on a description of a simplified axon. However as outlined above many more processes can be important and above extensions to take into account these effects, processes and structures have been outlined. Moreover, an axon itself is part of a neuron ending on one side with an axon hillock and on the other end with synapses. Another avenue of expansion is to include the adjacent structures by following a similar philosophy as has been used for setting up the present framework. It must be noted that this does not mean including these in the current framework as it is but rather means developing a separate model best suited for describing these (for example, for synapse and axon hillock and/or soma of the neuron) and then linking these up to the present framework through contact forces or initial conditions. In essence, this means by using simple enough building blocks in which the causal relationships between the described effects can be followed and then use these understandable building blocks to build up to the larger systems, establishing a trail of causes and effects which can be traced back to the basic physical principles even from larger and potentially highly complex levels. One could say this is an 'atomistic' modelling philosophy in which the axon model is made up of models describing the single effects composed into a whole by the added coupling forces and which, in turn, can be used as a building block for even larger structures and systems.

As it stands in the present framework, the concentration gradient across the biomembrane is taken as *a priori* granted. As an example, once the model extends to the whole neural cell in which the axon is only one part this concentration difference might no longer be a constant. Indeed, it might fall out from the biochemical description of the rest of the neuron and in practical terms, this could mean using a more accurate model for obtaining the AP signal capable of taking into account these time-dependent concentration changes.

In all modifications one should not forget the Occam's razor: "Entities should not be multiplied without necessity" (attributed to William of Ockham, 1287-1347). It means that a basic model should describe the basic effects and then it could be modified for specific, mostly pathological situations.

8.5 Dimensions

Up until now, we have been using dimensionless governing equations with corresponding dimensionless parameters. In reality, the dimensions must be taken into account. Although the numerical simulations in this book (see further in Chap. 9) are presented in the dimensionless form stressing the qualitative side of the modelling, the overview on dimensions must also be presented. Such an analysis is important for further experimental studies.

In the present Section, the lower case variables and parameters are used to emphasise the presentation with dimensions while the upper case variables and parameters are used for the dimensionless setting unless noted otherwise. Notations for dimensions are presented in square brackets [. . .] throughout the Section. Basic units of the SI system are preferred but for the sake of readability, some SI derived units are used (for example, potential is given in volts). The derived SI units used in this text (like Hz, H, F, Wb, etc.) will be summarised at the end of this Chapter.

The dimensionless variables Z, J, U, W, P and Θ used for modelling the wave ensemble (see Sect. 8.1) are denoted by z, j, u, w, p and θ in their physical units. The coupling forces are denoted F_i and f_i, respectively and the parameters in physical units are noted by subscript d.

The units have the following notations: [s] - second (time), [m] - metre (space), [kg] - kilogram (mass), [V] - volt (potential), [A] - ampere (current), [K] - Kelvin (temperature) and are used without SI prefixes, i.e., AP is given in [V] not in [mV] as measured in experiments.

8.5.1 Governing Equations in Physical Units

The telegraph equations

The initial telegraph equations are the basic model for all the AP models derived by Hodgkin and Huxley, Nagumo et al. These equations are [42]

$$-j_x = i + \pi a^2 C_a v_t,$$
$$-v_x = rj + L/(\pi a^2) j_t, \tag{8.19}$$

or

$$v_{xx} - LC_a v_{tt} = RC_a v_t + \frac{2}{a} RI + \frac{2}{a} LI_t, \tag{8.20}$$

where small subscripts t and x denote derivation with respect to time and space, respectively (before capital subscripts were used), t is time [s], x is distance [m], a is axon radius [m], v is potential difference [V], i is axon current per unit length [A/m], j is membrane current per unit length [A/m] (i_a in original notation), $I = i/(2\pi a)$ is axon current density [A/m²], r is resistance per unit length of axon [Ω/m], $R = (\pi a^2) \cdot r$ is specific resistance [$\Omega \cdot$ m], L is axon specific self-inductance [H] and C_a is axon self-capacitance per unit area per unit length [F/m²].

The HH model

Hodgkin and Huxley based their model [30] on the telegraph equations by noting that the inductance is small enough to be possible to neglect. The main part of the HH model is the expression for the membrane current:

$$I = C_0 V_t + g_{Na} m^3 h(V - V_{Na}) + g_K n^4 (V - V_K) + g_L (V - V_L), \tag{8.21}$$

where sub-index t denotes derivative with respect to time, t is time [s], I is membrane current density [A/m²], V is membrane voltage [V], C_0 is membrane capacity [F/m²], g_{Na}, g_K, g_L are coefficients [Ω/m²] and V_{Na}, V_K, V_L are potential change contributions from individual ion currents [V]. The cable model describing the propagating pulse can be written as a PDE also [30] (which is the version used for the numerical example given in the present book, see Fig. 6.1)

$$CV_t = \frac{a}{2R}V_{xx} + I - I_K - I_{Na} - I_L, \qquad (8.22)$$

where sub-index x denotes spatial partial derivative, sub-index t denotes temporal partial derivative, t is time [s], x is distance [m], C is membrane capacity [F/m²], V is membrane voltage [V], I is the current density [A/m²], a is axon radius [m] and R is the specific resistance of the axoplasm [Ω/m]. As the HH model is commonly used for interpreting experimental observations of the APs, it is normally used in its dimensional form [5, 8, 37].

The FHN model

The FHN model (see Sect. 6.1.2) is a further simplification of the HH model where all the individual ion currents have been replaced by one current only [50]. However, FitzHugh [22] named this model after Bonhoeffer and van der Pol (BVP) and presented it in the dimensionless form. He stated: "The BVP model is not intended to be an accurate quantitative model of the axon, in the sense of reproducing the shape of experimental curves". He took this model as a representative of a wider class of non-linear excitable-oscillatory systems. From the viewpoint of an axon, this model shows a typical shape of the AP, the threshold and refractory properties and its phase-plane analysis has demonstrated the richness of dynamics described by it [3, 22, 52]. Using the BVP model as a base, several types of oscillations can be described by its modifications [41]. McKean [46] has noted that a general form of the HH model is

$$e_t = e_{xx} + f, \qquad (8.23)$$

where subscript t denotes partial derivative with respect to time and x partial derivative with respect to space. In comparison the FHN model can be written as

$$e_t = e_{xx} + f - b \int e \, dt, \qquad (8.24)$$

in which f is cubic polynomial, b is positive constant and $\int e \, dt$ is an indefinite integral. Usually the FHN model is represented in a dimensionless form [22, 50]. However, Bini et al. [4] started from a rather general form in physical units, including also the temperature effects. Their model (in original notations) is the following:

$$i_t = \Phi(T)\left(\frac{V - V_0 - Ri}{L}\right), \qquad (8.25)$$

$$C_m V_t - D_0 \nabla^2 V = -\left(\frac{\eta(T)}{A}\right)[F(V) + i] - I_0, \tag{8.26}$$

$$k_0 \nabla^2 T + \sigma_0 (\nabla V)^2 + c_* w_* (T_* - T) = \rho c_p \partial_t T, \tag{8.27}$$

where V is the amplitude of AP, i is a gating variable, T is the temperature and the rest of the variables belong to the circuit equivalent representation of the FHN model or characterise heat conduction properties of biological tissues (see Appendix A in [4] for details). In the dimensionless form (see Appendix B in [4] for details) Eqs. (8.25)–(8.27) can be written as

$$v_\tau = \frac{1}{\chi}\left\{D_1 \nabla^2 v + (1 + b\Theta)\left[\Sigma v(1 - v)(v - a) - w\right] - w_0\right\},$$

$$w_\tau = \frac{1}{\chi} 3^\Theta \epsilon_{21}(v - v_0 - \gamma w), \tag{8.28}$$

$$\Theta_\tau = \frac{1}{\chi}\left[\epsilon_{24} \nabla^2 \Theta + \epsilon_{25} \sigma_1 \nabla v \cdot \nabla v - \epsilon_{23}(\Theta - \Theta_*)\right].$$

In Eq. (8.28) τ denotes dimensionless time, χ dimensionless space, v dimensionless AP, w is gating variable, Σ is the free parameter and Θ represents dimensionless temperature. If in Eq. (8.28) the temperature Θ is taken constant then one can get the FHN model where the temperature dependence of coefficients is taken into account:

$$v_\tau = \frac{1}{\chi}\left\{D_1 \nabla^2 v + (1 + b\Theta)\left[\Sigma v(1 - v)(v - a) - w\right] - w_0\right\},$$

$$w_\tau = \frac{1}{\chi} 3^\Theta \epsilon_{21}(v - v_0 - \gamma w) \tag{8.29}$$

and if $w_0 = v_0 = 0$ then our model (8.1) is easily recognised.

Clearly, one can use for the FHN model with dimensions after Bini et al. [4]. However, following Eqs. (8.1), the equations with dimensions (for z [V] and j [A]) could be rewritten as

$$z_t = d z_{xx} - \zeta_{1d}(a_1 + b_1) z + \zeta_{2d}(1 + (a_1 + b_1)) z^2 - \zeta_{3d} z^3 - \iota_{1d} j,$$

$$j_t = \epsilon \zeta_{4d}(a_2 + b_2) z - \epsilon \iota_{2d} j. \tag{8.30}$$

Dimensions of the variables and their derivatives are following:

$$z\,[V], \quad z_t\left[\frac{V}{s}\right], \quad z_{xx}\left[\frac{V}{m^2}\right], \quad j_t\left[\frac{A}{s}\right]. \tag{8.31}$$

Parameters in Eq. (8.30) have following dimensions:

$$d\left[\frac{m^2}{s}\right], \quad \zeta_{1d}\left[\frac{1}{s}\right], \quad \zeta_{2d}\left[\frac{s^2 \cdot A}{kg \cdot m^2}\right], \quad \zeta_{3d}\left[\frac{s^5 \cdot A^2}{kg^2 \cdot m^4}\right], \quad \zeta_{4d}\left[\frac{s^2 \cdot A^2}{kg \cdot m^2}\right],$$

$$\iota_{1d}\left[\frac{kg \cdot m^2}{s^4 \cdot A^2}\right], \quad \iota_{2d}\left[\frac{1}{s}\right]. \tag{8.32}$$

The activation coefficients a_i and b_i and the time-scale coefficient ε are taken dimensionless. Noting that in (8.30) $b_1 = -\beta_1 u$ and $b_2 = -\beta_2 u$ where u is the LW from (8.37) with dimension $\left[\text{kg/m}^2\right]$ means that β_i has dimension $\left[\text{m}^2/\text{kg}\right]$.

In numerical calculations in physical units, it might be better to use the HH model because there exist many experiments and the needed set of parameters is clearly determined [8, 35, 37, 51].

The pressure changes

As the pressure change $p\left[\text{kg}/(\text{m} \cdot \text{s}^2)\right]$ is represented by a classical wave equation it is easy to switch between the dimensionless and dimensional versions

$$p_{tt} = c_2^2 p_{xx} - \mu_{2d} p_t + f_2(z, j), \tag{8.33}$$

where the dimensions are following:

$$\begin{aligned}
& p_t \left[\frac{\text{kg}}{\text{m} \cdot \text{s}^3}\right], \quad p_{tt} \left[\frac{\text{kg}}{\text{m} \cdot \text{s}^4}\right], \quad p_{xx} \left[\frac{\text{kg}}{\text{m}^3 \cdot \text{s}^2}\right], \\
& c_2^2 \left[\frac{\text{m}^2}{\text{s}^2}\right], \quad \mu_{2d} \left[\frac{1}{\text{s}}\right], \quad f_2(z, j) \left[\frac{\text{kg}}{\text{m} \cdot \text{s}^4}\right].
\end{aligned} \tag{8.34}$$

The improved Heimburg-Jackson model

The iHJ model starts with dimensions in the original notation [27, 28] and is then converted into the dimensionless form [16, 53] which is used for the numerical examples shown in Chap. 9. In the original notations [27] the HJ model can be written as

$$(\Delta\rho^A)_{tt} = \left[\left(c_0^2 + p\Delta\rho^A + q(\Delta\rho^A)^2\right)_x \Delta\rho^A\right]_x - h(\Delta\rho^A)_{xxxx}, \tag{8.35}$$

where subindices x and t as before, denote partial derivatives, $\Delta\rho^A$ is the change of the lateral density of the biomembrane in $[\text{kg/m}^2]$, c_0 is sound velocity in unperturbed state in [m/s], p is an nonlinear polynomial fitting coefficient for experimental data in $[c_0^2/\rho_0^A]$, q is an nonlinear polynomial fitting coefficient for experimental data in $[c_0^2/(\rho_0^A)^2]$ and h is *ad hoc* dispersion coefficient in $[\text{m}^4/\text{s}^2]$. The original HJ model (8.35) is not causal as it allows higher frequencies to travel at infinite velocity and was improved [16] by taking into account also the inertial properties of the biomembrane by including a second dispersive term $h_2 u_{xxtt}$ (where $u = \Delta\rho^A$ and h_2 is second dispersion coefficient in $[\text{m}^2]$) resulting in the iHJ model in dimensional form

$$u_{tt} = \left(c_{3d}^2 + pu + qu^2\right) u_{xx} + (p + 2qu)(u_x)^2 - h u_{xxxx} + h_2 u_{xxtt}, \tag{8.36}$$

where the dimensions are the same as noted for the (8.35) with $c_{3d} = c_0$. Including the coupling force and changing the notations for the nonlinear parameters the iHJ model for the density change $u \left[\text{kg/m}^2\right]$ can be written as

$$u_{tt} = c_{3d}^2 u_{xx} + nuu_{xx} + mu^2 u_{xx} + nu_x^2 + 2muu_x^2 -$$
$$- h_1 u_{xxxx} + h_2 u_{xxtt} - \mu_{3d} u_t + f_3(z, j, p), \tag{8.37}$$

with the following dimensions of the derivatives:

$$u_{tt} \left[\frac{\text{kg}}{\text{m}^2 \cdot \text{s}^2}\right], \quad u_{xx} \left[\frac{\text{kg}}{\text{m}^4}\right], \quad u_{xxxx} \left[\frac{\text{kg}}{\text{m}^6}\right], \quad u_{xxtt} \left[\frac{\text{kg}}{\text{m}^4 \cdot \text{s}^2}\right],$$
$$u_t \left[\frac{\text{kg}}{\text{m}^2 \cdot \text{s}}\right], \quad u^2 \left[\frac{\text{kg}^2}{\text{m}^4}\right], \quad u_x^2 \left[\frac{\text{kg}^2}{\text{m}^6}\right]. \tag{8.38}$$

The dimensions of the parameters are following:

$$c_{3d}^2 \left[\frac{\text{m}^2}{\text{s}^2}\right], \quad n \left[\frac{\text{m}^4}{\text{g} \cdot \text{s}^2}\right], \quad m \left[\frac{\text{m}^6}{\text{g}^2 \cdot \text{s}^2}\right], \quad h_1 \left[\frac{\text{m}^4}{\text{s}^2}\right], \quad h_2 \left[\text{m}^2\right],$$
$$\mu_{3d} \left[\frac{1}{\text{s}}\right], \quad f_3(z, j, p) \left[\frac{\text{kg}}{\text{m}^2 \cdot \text{s}^2}\right]. \tag{8.39}$$

and nonlinear coefficients are $n = p$ and $m = q$ compared to Eqs. (8.35) and (8.36) to keep in line with the notations used elsewhere (see Eq. (8.3)).

For the dimensionless case

$$X = \frac{x}{l}, \ T = \frac{c_0 t}{l}, \ U = \frac{u}{\rho_0}, \ P = p\frac{\rho_0}{c_0^2}, \ Q = q\frac{\rho_0^2}{c_0^2}, \ H_1 = \frac{h}{c_0^2 l^2}, \ H_2 = \frac{h_2}{l^2}, \tag{8.40}$$

where l is a certain length (for example, radius of the axon), x is space, t is time, capital letters are used to emphasise that the relevant parameter is dimensionless and the dimensions for the other quantities are the same as in Eqs. (8.35) and (8.36).

Transverse displacement

The transverse displacement TW is calculated like in the theory of rods:

$$w = ku_x, \tag{8.41}$$

where k is a coefficient; in theory of rods it is Poissons's ratio v. Here the dimensions of the variable w an derivative u_x are

$$w \ [\text{m}], \quad u_x \left[\frac{\text{kg}}{\text{m}^3}\right]. \tag{8.42}$$

The coefficient k has the following dimension:

$$k \left[\frac{\text{m}^4}{\text{kg}}\right]. \tag{8.43}$$

The heat equation

The heat equation is one of the classical equations of physics and is trivial to switch between the dimensionless (variable Θ) and dimensional (variable θ) forms, however, for the sake of completeness it is included so θ [K] can be written as

$$\theta_t = \alpha_d \theta_{xx} + f_4(z, j, u, p), \tag{8.44}$$

with the following dimensions of the derivatives

$$\theta_t \left[\frac{K}{s}\right], \quad \theta_{xx} \left[\frac{K}{m^2}\right] \tag{8.45}$$

and parameters

$$\alpha_d \left[\frac{m^2}{s}\right], \quad f_4(z, j, u, p) \left[\frac{K}{s}\right]. \tag{8.46}$$

8.5.2 Coupling Forces in Physical Units

Coupling forces are the important elements of the model which describe the interaction between the components of a wave ensemble and are responsible for the energy transfer and synchronisation. If the AP is described by a dimensionless FHN model, then coupling coefficients or parameters, like b_i are dimensionless. If the HH model is used then the feedback from the biomembrane dynamics is also with certain dimensions specified by dimensional analysis.

The coupling force for the pressure

The coupling force f_2 in Eq. (8.33) is

$$f_2 = \eta_{1d} z_x + \eta_{2d} j_t + \eta_{3d} z_t, \tag{8.47}$$

with the following dimensions of the derivatives

$$z_x \left[\frac{V}{m}\right], \quad j_t \left[\frac{A}{s}\right], \quad z_t \left[\frac{V}{s}\right] \tag{8.48}$$

and parameters

$$\eta_{1d} \left[\frac{A}{m^2 \cdot s}\right], \quad \eta_{2d} \left[\frac{kg}{m \cdot s^3 \cdot A}\right], \quad \eta_{3d} \left[\frac{A}{m^3}\right], \quad f_2 \left[\frac{kg}{m \cdot s^4}\right]. \tag{8.49}$$

The coupling force for the iHJ equation

The coupling force f_3 in Eq. (8.37) is

$$f_3 = \gamma_{1d} p_t + \gamma_{2d} j_t - \gamma_{3d} z_t, \tag{8.50}$$

with the following dimensions of the derivatives

$$p_t \left[\frac{kg}{m \cdot s^3}\right], \quad j_t \left[\frac{A}{s}\right], \quad z_t \left[\frac{V}{s}\right] \tag{8.51}$$

and parameters

$$\gamma_{1d} \left[\frac{s}{m}\right], \quad \gamma_{2d} \left[\frac{kg}{m^2 \cdot s \cdot A}\right], \quad \gamma_{3d} \left[\frac{s^2 \cdot A}{m^4}\right], \quad f_3 \left[\frac{kg}{m^2 \cdot s^2}\right]. \tag{8.52}$$

The coupling force for the heat equation

From Eq. (8.44) one can see that the corresponding coupling force f_4 is

$$f_4 = \tau_{1d} z^2 + \tau_{2d} (p_t + \varphi_{2d}(p)) + \tau_{3d} (u_t + \varphi_{3d}(u)) - \tau_{4d} \omega, \tag{8.53}$$

with the following dimensions of the variable z and derivatives

$$z \, [V], \quad p_t \left[\frac{kg}{m \cdot s^3}\right], \quad u_t \left[\frac{kg}{m^2 \cdot s}\right]. \tag{8.54}$$

Dimensions of the parameters are

$$\tau_{1d} \left[\frac{s^5 \cdot A^2 \cdot K}{kg^2 \cdot m^4}\right], \quad \tau_{2d} \left[\frac{m \cdot s^2 \cdot K}{kg}\right], \quad \tau_{3d} \left[\frac{m^2 \cdot K}{kg}\right], \quad \tau_{4d} \, [s], \quad \omega \left[\frac{K}{s^2}\right],$$
$$\varphi_{2d}(p) \left[\frac{kg}{m \cdot s^3}\right], \quad \varphi_{3d}(u) \left[\frac{kg}{m^2 \cdot s}\right], \quad f_4 \left[\frac{K}{s}\right]. \tag{8.55}$$

The dimensional expressions of φ_{2d} and φ_{3d} in (8.53) can be written (for φ_{4d} see Eq. (8.11)) as

$$\varphi_{2d}(p) = \lambda_{2d} \int p_t dt,$$
$$\varphi_{3d}(u) = \lambda_{3d} \int u_t dt, \tag{8.56}$$
$$\varphi_{4d}(J) = \zeta_d \int j dt,$$

with following dimensions of integrals

$$\int p_t dt \left[\frac{kg}{m \cdot s^2}\right], \quad \int u_t dt \left[\frac{kg}{m^2}\right], \quad \int j dt \, [A \cdot s] \tag{8.57}$$

and parameters

$$\lambda_{2d} \left[\frac{1}{s}\right], \quad \lambda_{3d} \left[\frac{1}{s}\right], \quad \zeta_d \left[\frac{K}{A \cdot s^3}\right]. \tag{8.58}$$

8.5.3 Summary

The parameters and their dimensions are summarised in Table 8.1.

Table 8.1 Parameters and their dimensions.

AP Eq. (8.30)									
Parameter	d	ζ_{1d}	ζ_{2d}	ζ_{3d}	ζ_{d4}	ι_{1d}	ι_{2d}	β_1	β_2
Dimension	$\left[\frac{m^2}{s}\right]$	$\left[\frac{1}{s}\right]$	$\left[\frac{s^2 \cdot A}{kg \cdot m^2}\right]$	$\left[\frac{s^5 \cdot A}{kg^2 \cdot m^4}\right]$	$\left[\frac{s^2 \cdot A^2}{kg \cdot m^2}\right]$	$\left[\frac{kg \cdot m^2}{s^4 \cdot A^2}\right]$	$\left[\frac{1}{s}\right]$	$\left[\frac{m^2}{kg}\right]$	$\left[\frac{m^2}{kg}\right]$
PW Eq. (8.33)									
Parameter	c_f	μ_{2d}	η_{1d}	η_{2d}	η_{3d}				
Dimension	$\left[\frac{m}{s}\right]$	$\left[\frac{1}{s}\right]$	$\left[\frac{A}{m^2 \cdot s}\right]$	$\left[\frac{kg}{m \cdot s^3 \cdot A}\right]$	$\left[\frac{A}{m^3}\right]$				
LW Eq. (8.37)									
Parameter	c_{3d}	μ_{3d}	n	m	h_1	h_2	γ_{1d}	γ_{2d}	γ_{3d}
Dimension	$\left[\frac{m}{s}\right]$	$\left[\frac{1}{s}\right]$	$\left[\frac{m^4}{kg \cdot s^2}\right]$	$\left[\frac{m^6}{kg^2 \cdot s^2}\right]$	$\left[\frac{m^4}{s^2}\right]$	$[m^2]$	$\left[\frac{s}{m}\right]$	$\left[\frac{kg}{m^2 \cdot s \cdot A}\right]$	$\left[\frac{s^2 \cdot A}{m^4}\right]$
TW Eq. (8.41)									
Parameter	k								
Dimension	$\left[\frac{m^4}{kg}\right]$								
θ Eq. (8.44)									
Parameter	α_d	λ_{2d}	λ_{3d}	τ_{1d}	τ_{2d}	τ_{3d}	τ_{4d}		
Dimension	$\left[\frac{m^2}{s}\right]$	$\left[\frac{1}{s}\right]$	$\left[\frac{1}{s}\right]$	$\left[\frac{s^5 \cdot A^2 \cdot K}{kg \cdot m^4}\right]$	$\left[\frac{m \cdot s^2 \cdot K}{kg}\right]$	$\left[\frac{K \cdot m^2}{kg}\right]$	$[s]$		

The following definitions of SI system units could be useful for describing some parameters

$$[Pa] = \left[\frac{kg}{m \cdot s^2}\right], \quad [Hz] = \left[\frac{1}{s}\right], \quad [H] = \left[\frac{kg \cdot m^2}{s^2 \cdot A}\right], \quad [J] = \left[\frac{kg \cdot m^2}{s^2}\right],$$

$$[N] = \left[\frac{kg \cdot m}{s^2}\right], \quad [F] = \left[\frac{s^4 \cdot A^2}{kg \cdot m^2}\right], \quad [Wb] = \left[\frac{kg \cdot m^2}{s^2 \cdot A}\right], \quad [C] = [A \cdot s], \tag{8.59}$$

where [Pa] is Pascal (pressure), [Hz] is Hertz (frequency), [H] is Henry (inductance), [J] is Joule (energy), [N] is Newton (force), [F] is Farad (capacitance), [Wb] is Weber (magnetic flux) and [C] is Coulomb (charge), although for the sake of clarity mostly base units are used.

References

1. Abbott, B.C., Hill, A.V., Howarth, J.V.: The positive and negative heat production associated with a nerve impulse. Proc. R. Soc. B Biol. Sci. **148**(931), 149–187 (1958). DOI 10.1098/rspb.1958.0012

2. Barz, H., Schreiber, A., Barz, U.: Impulses and pressure waves cause excitement and conduction in the nervous system. Med. Hypotheses **81**(5), 768–772 (2013). DOI 10.1016/j.mehy.2013.07.049

3. Binczak, S., Jacquir, S., Bilbault, J.M., Kazantsev, V.B., Nekorkin, V.I.: Experimental study of electrical FitzHugh–Nagumo neurons with modified excitability. Neural Networks **19**(5), 684–693 (2006). DOI 10.1016/j.neunet.2005.07.011.

4. Bini, D., Cherubini, C., Filippi, S.: Heat transfer in Fitzhugh-Nagumo models. Phys. Rev. E **74**(4), 041905 (2006). DOI 10.1103/PhysRevE.74.041905.

5. Chen, H., Garcia-Gonzalez, D., Jérusalem, A.: Computational model of the mechanoelectrophysiological coupling in axons with application to neuromodulation. Phys. Rev. E **99**(3), 032406 (2019). DOI 10.1103/PhysRevE.99.032406

6. Christov, C.I., Maugin, G.A., Porubov, A.V.: On Boussinesq's paradigm in nonlinear wave propagation. Comptes Rendus Mécanique **335**(9–10), 521–535 (2007). DOI 10.1016/j.crme.2007.08.006

7. Christov, C.I., Velarde, M.G.: Dissipative solitons. Phys. D Nonlinear Phenom. **86**(1–2), 323–347 (1995). DOI 10.1016/0167-2789(95)00111-G

8. Courtemanche, M., Ramirez, R.J., Nattel, S.: Ionic mechanisms underlying human atrial action potential properties : insights from a mathematical model. Am. J. Physiol. **275**(1), 301–321 (1998)

9. Debanne, D., Campanac, E., Bialowas, A., Carlier, E., Alcaraz, G.: Axon physiology. Physiol. Rev. **91**(2), 555–602 (2011). DOI 10.1152/physrev.00048.2009

10. Deseri, L., Piccioni, M.D., Zurlo, G.: Derivation of a new free energy for biological membranes. Contin. Mech. Thermodyn. **20**(5), 255–273 (2008). DOI 10.1007/s00161-008-0081-1

11. Drukarch, B., Holland, H.A., Velichkov, M., Geurts, J.J., Voorn, P., Glas, G., de Regt, H.W.: Thinking about the nerve impulse: A critical analysis of the electricity-centered conception of nerve excitability. Prog. Neurobiol. **169**, 172–185 (2018). DOI 10.1016/j.pneurobio.2018.06.009

12. El Hady, A., Machta, B.B.: Mechanical surface waves accompany action potential propagation. Nat. Commun. **6**, 6697 (2015). DOI 10.1038/ncomms7697

13. Engelbrecht, J.: On theory of pulse transmission in a nerve fibre. Proc. R. Soc. A Math. Phys. Eng. Sci. **375**(1761), 195–209 (1981). DOI 10.1098/rspa.1981.0047.

14. Engelbrecht, J., Peets, T., Tamm, K.: Electromechanical coupling of waves in nerve fibres. Biomech. Model. Mechanobiol. **17**(6), 1771–1783 (2018). DOI 10.1007/s10237-018-1055-2

15. Engelbrecht, J., Peets, T., Tamm, K., Laasmaa, M., Vendelin, M.: On the complexity of signal propagation in nerve fibres. Proc. Estonian Acad. Sci. **67**(1), 28–38 (2018). DOI 10.3176/proc.2017.4.28

16. Engelbrecht, J., Tamm, K., Peets, T.: On mathematical modelling of solitary pulses in cylindrical biomembranes. Biomech. Model. Mechanobiol. **14**(1), 159–167 (2015). DOI 10.1007/s10237-014-0596-2

17. Engelbrecht, J., Tamm, K., Peets, T.: Modeling of complex signals in nerve fibers. Med. Hypotheses **120**, 90–95 (2018). DOI 10.1016/j.mehy.2018.08.021

18. Engelbrecht, J., Tamm, K., Peets, T.: Criteria for modelling wave phenomena in complex systems: the case of signals in nerves. Proc. Estonian Acad. Sci. **68**(3), 276 (2019). DOI 10.3176/proc.2019.3.05

19. Engelbrecht, J., Tamm, K., Peets, T.: Internal variables used for describing the signal propagation in axons. Contin. Mech. Thermodyn. **32**(6), 1619–1627 (2020). DOI 10.1007/s00161-020-00868-2

20. Engelbrecht, J., Tamm, K., Peets, T.: On mechanisms of electromechanophysiological interactions between the components of nerve signals in axons. Proc. Estonian Acad. Sci. **69**(2), 81–96 (2020). DOI 10.3176/proc.2020.2.03

21. Fillafer, C., Mussel, M., Muchowski, J., Schneider, M.F.: Cell surface deformation during an action potential. Biophys. J. **114**(2), 410–418 (2018). DOI 10.1016/j.bpj.2017.11.3776

22. FitzHugh, R.: Impulses and physiological states in theoretical models of nerve membrane. Biophys. J. **1**(6), 445–466 (1961). DOI 10.1016/S0006-3495(61)86902-6

23. Græsbøll, K., Sasse-Middelhoff, H., Heimburg, T.: The thermodynamics of general and local anesthesia. Biophys. J. **106**(10), 2143–2156 (2014). DOI 10.1016/j.bpj.2014.04.014.

24. Gross, D., Williams, W.S., Connor, J.A.: Theory of electromechanical effects in nerve. Cell. Mol. Neurobiol. **3**(2), 89–111 (1983). DOI 10.1007/BF00735275

25. Heimburg, T.: Lipid ion channels.   Biophys. Chem. **150**(1-3), 2–22 (2010).   DOI 10.1016/j.bpc.2010.02.018.

26. Heimburg, T.: The important consequences of the reversible heat production in nerves and the adiabaticity of the action potential. arXiv:2002.06031 [physics.bio-ph] (2020)

27. Heimburg, T., Jackson, A.D.: On soliton propagation in biomembranes and nerves. Proc. Natl. Acad. Sci. USA **102**(28), 9790–9795 (2005). DOI 10.1073/pnas.0503823102

28. Heimburg, T., Jackson, A.D.: On the action potential as a propagating density pulse and the role of anesthetics. Biophys. Rev. Lett. **02**(01), 57–78 (2007). DOI 10.1142/S179304800700043X

29. Heimburg, T., Jackson, A.D.: Thermodynamics of the nervous impulse. In: N. Kaushik (ed.) Structure and dynamics of membranous interfaces, chap. 12, pp. 318–337. John Wiley & Sons (2008)

30. Hodgkin, A.L., Huxley, A.F.: A quantitative description of membrane current and its application to conduction and excitation in nerve. J. Physiol. **117**(4), 500–544 (1952). DOI 10.1113/jphysiol.1952.sp004764

31. Hodgkin, A.L., Katz, B.: The effect of temperature on the electrical activity of the giant axon of the squid. J. Physiol. **109**(1-2), 240–249 (1949). DOI 10.1113/jphysiol.1949.sp004388.

32. Howarth, J.V., Keynes, R.D., Ritchie, J.M.: The origin of the initial heat associated with a single impulse in mammalian non-myelinated nerve fibres. J. Physiol. **194**(3), 745–93 (1968). DOI 10.1113/jphysiol.1968.sp008434

33. Howarth, J.V., Keynes, R.D., Ritchie, J.M., von Muralt, A.: The heat production associated with the passage of a single impulse in pike olfactory nerve fibres. J. Physiol. **249**(2), 349–368 (1975). DOI 10.1113/jphysiol.1975.sp011019

34. Iwasa, K., Tasaki, I., Gibbons, R.: Swelling of nerve fibers associated with action potentials. Science **210**(4467), 338–339 (1980). DOI 10.1126/science.7423196

35. Izhikevich, E.M.: Dynamical Systems in Neuroscience: The Geometry of Excitability and Bursting. MIT Press, London (2005)

36. Jerusalem, A., Al-Rekabi, Z., Chen, H., Ercole, A., Malboubi, M., Tamayo-Elizalde, M., Verhagen, L., Contera, S.: Electrophysiological-mechanical coupling in the neuronal membrane and its role in ultrasound neuromodulation and general anaesthesia. Acta Biomater. **97**, 116–140 (2019). DOI 10.1016/j.actbio.2019.07.041.

37. Jérusalem, A., García-Grajales, J.A., Merchán-Pérez, A., Peña, J.M.: A computational model coupling mechanics and electrophysiology in spinal cord injury. Biomech. Model. Mechanobiol. **13**(4), 883–896 (2014). DOI 10.1007/s10237-013-0543-7

38. Kang, K.H., Schneider, M.F.: Nonlinear pulses at the interface and its relation to state and temperature. Eur. Phys. J. E **43**(2), 8 (2020). DOI 10.1140/epje/i2020-11903-x.

39. Kappler, J., Shrivastava, S., Schneider, M.F., Netz, R.R.: Nonlinear fractional waves at elastic interfaces. Phys. Rev. Fluids **2**, 114804 (2017). DOI 10.1103/PhysRevFluids.2.114804

40. Kaufmann, K.: Action Potentials and Electromechanical Coupling in the Macroscopic Chiral Phospholipid Bilayer. Caruaru (1989)

41. Kutafina, E.: Mixed mode oscillations in the Bonhoeffer-van der Pol oscillator with weak periodic perturbation. Comput. Appl. Math. **34**(1), 81–92 (2015). DOI 10.1007/s40314-013-0105-8

42. Lieberstein, H.: On the Hodgkin-Huxley partial differential equation. Math. Biosci. **1**(1), 45–69 (1967). DOI 10.1016/0025-5564(67)90026-0.

43. Lucassen, J., van der Tempel, M.: Longitudinal waves on viscoelastic surfaces. J. Colloid Interface Sci. **41**(3), 491–498 (1972). DOI 10.1016/0021-9797(72)90373-6

44. Lundström, I.: Mechanical wave propagation on nerve axons. J. theor. Biol. **45**, 487–499 (1974)
45. Margineanu, D.G., Schoffeniels, E.: Molecular events and energy changes during the action potential. Proc. Natl. Acad. Sci. **74**(9), 3810–3813 (1977). DOI 10.1073/pnas.74.9.3810.
46. McKean, H.: Nagumo's equation.   Adv. Math. (N. Y). **4**(3), 209–223 (1970).   DOI 10.1016/0001-8708(70)90023-X.
47. Meissner, S.T.: Proposed tests of the soliton wave model of action potentials, and of inducible lipid pores, and how non-electrical phenomena might be consistent with the Hodgkin-Huxley model. arXiv:1808.07193 [physics.bio-ph] (2018)
48. Morris, C., Lecar, H.: Voltage oscillations in the barnacle giant muscle fiber. Biophys. J. **35**(1), 193–213 (1981). DOI 10.1016/S0006-3495(81)84782-0
49. Mueller, J.K., Tyler, W.J.: A quantitative overview of biophysical forces impinging on neural function. Phys. Biol. **11**(5), 051001 (2014). DOI 10.1088/1478-3975/11/5/051001
50. Nagumo, J., Arimoto, S., Yoshizawa, S.: An active pulse transmission line simulating nerve axon. Proc. IRE **50**(10), 2061–2070 (1962). DOI 10.1109/JRPROC.1962.288235
51. Nelson, P.C., Radosavljevic, M., Bromberg, S.: Biological Physics: Energy, Information, Life. W.H. Freeman and Company, New York, NY (2003)
52. Neu, J.C., Preissig, R., Krassowska, W.: Initiation of propagation in a one-dimensional excitable medium. Phys. D Nonlinear Phenom. **102**(3-4), 285–299 (1997). DOI 10.1016/S0167-2789(96)00203-5
53. Peets, T., Tamm, K.: On mechanical aspects of nerve pulse propagation and the Boussinesq paradigm. Proc. Estonian Acad. Sci. **64**(3), 331 (2015). DOI 10.3176/proc.2015.3S.02
54. Petrov, A.G.: Flexoelectricity of model and living membranes. Biochim. Biophys. Acta - Biomembr. **1561**(1), 1–25 (2002). DOI 10.1016/S0304-4157(01)00007-7.
55. Petrov, A.G.: Electricity and mechanics of biomembrane systems: Flexoelectricity in living membranes. Anal. Chim. Acta **568**(1-2), 70–83 (2006). DOI 10.1016/j.aca.2006.01.108
56. Porubov, A.V.: Amplification of Nonlinear Strain Waves in Solids. World Scientific, Singapore (2003)
57. Ritchie, J.M., Keynes, R.D.: The production and absorption of heat associated with electrical activity in nerve and electric organ.   Q. Rev. Biophys. **18**(04), 451 (1985).   DOI 10.1017/S0033583500005382
58. Scott, A.C.: Nonlinear Science. Emergence and Dynamics of Coherent Structures. Oxford University Press (1999)
59. Tamm, K., Engelbrecht, J., Peets, T.: Temperature changes accompanying signal propagation in axons. J. Non-Equilibrium Thermodyn. **44**(3), 277–284 (2019). DOI 10.1515/jnet-2019-0012
60. Tasaki, I.: A macromolecular approach to excitation phenomena: mechanical and thermal changes in nerve during excitation. Physiol. Chem. Phys. Med. NMR **20**(4), 251–268 (1988)
61. Tasaki, I., Fujita, M.: Action currents of single nerve fibers as modified by temperature changes. J. Neurophysiol. **11**(4), 311–315 (1948). DOI 10.1152/jn.1948.11.4.311.
62. Terakawa, S.: Potential-dependent variations of the intracellular pressure in the intracellularly perfused squid giant axon.   J. Physiol. **369**(1), 229–248 (1985).   DOI 10.1113/jphysiol.1985.sp015898
63. Yang, Y., Liu, X.W., Wang, H., Yu, H., Guan, Y., Wang, S., Tao, N.: Imaging action potential in single mammalian neurons by tracking the accompanying sub-nanometer mechanical motion. ACS Nano **12**(5), 4186–4193 (2018). DOI 10.1021/acsnano.8b00867.

Chapter 9
In Silico Experiments

> *Equations are the lifeblood of mathematics, science and technology.*
>
> *Ian Stewart, 2013*

The full coupled model for describing an ensemble of waves in a nerve axon is presented in Sect. 8.1 (see Eqs. (8.1)-(8.11) and Fig. 8.1). In this Chapter, the proposed model is analysed by numerical simulation. This concerns solving a system of partial differential equations which describes more or less the important physical effects related to the propagation of the AP, longitudinal wave LW in the biomembrane, pressure wave in the axoplasm PW, transverse displacement of the biomembrane TW, and temperature change Θ, which all are coupled. We start with simplified variants of system (8.1)-(8.11) and conclude with numerical analysis of the full system [5, 8, 28].

9.1 Numerical Method

The analysis of the proposed model is carried out by solving the system of partial differential equations (PDE) (8.1)-(8.11) numerically by using the discrete Fourier transform (DFT) based pseudospectral method (PSM) [27]. The idea of the PSM is to write the PDEs in a form where all the time derivatives are on the left hand side and all the spatial derivatives on the right hand side of the equation, and then apply properties of the Fourier transform for finding the spatial derivatives. This approach makes it possible to reduce the PDE into an ordinary differential equation (ODE) which can be solved by standard numerical methods. The full details of the numerical scheme are given in Appendix A and and example script in Appendix B.

For the numerical solving, a sech^2-type localised initial condition is used for Z:

$$Z(X,0) = Z_0 \, \mathrm{sech}^2 B_0 X, \qquad (9.1)$$

where Z_0 is the initial amplitude of the AP and B_0 is the initial pulse width. This initial 'spark' is taken above the threshold value in the middle of the space domain causing the AP to emerge and propagate to the left and right. All other processes have zero initial conditions and are generated by the propagating AP as a result of

J. Engelbrecht et al., *Modelling of Complex Signals in Nerves*,
https://doi.org/10.1007/978-3-030-75039-8_9

coupling terms added to the system. Boundary conditions are taken periodic as is required for the PSM. The time integration intervals have been taken short enough to avoid interaction between counter-propagating waves at the boundaries. In the figures only left propagating waves are plotted.

9.2 AP Coupled to Single Mechanical Effects

When the AP is coupled to the mechanical effects only then the system (8.1)-(8.11) is reduced to following set of equations [5, 8]:

$$Z_T = DZ_{XX} - J + Z\left(Z - C_1 - Z^2 + C_1 Z\right),$$
$$J_T = \epsilon_1 \left(C_2 Z - J\right), \tag{9.2}$$

$$P_{TT} = c_2^2 P_{XX} - \mu_2 P_T + F_2(Z, J), \tag{9.3}$$

$$U_{TT} = c_3^2 U_{XX} + N U U_{XX} + M U^2 U_{XX} + N U_X^2 + 2 M U U_X^2$$
$$- H_1 U_{XXXX} + H_2 U_{XXTT} + F_3(Z, J, P), \tag{9.4}$$

$$W = K U_X, \tag{9.5}$$

$$F_2 = \eta_1 Z_X + \eta_2 J_T, \tag{9.6}$$

$$F_3 = \gamma_1 P_T + \gamma_2 J_T, \tag{9.7}$$

where the AP is governed by the FHN model (9.2), the PW is governed by the modified wave equation (9.3), the LW is governed by the improved HJ model (9.4) and the TW is calculated like in the theory of rods – by Eq. (9.5). The coupling between the AP and PW is modelled by Eq. (9.6) and between the AP and LW by Eq. (9.7). Note that comapared to Eqs. (8.3), (8.6) and (8.7), here $\mu_3 = \eta_3 = \gamma_3 = 0$.

One possible physical interpretation of the proposed terms in Eqs. (9.6) and (9.7) is that the time derivatives could be interpreted as forces acting across the lipid bi-layer at a fixed spatial point on axon while spatial gradients could be interpreted as forces acting along the axon axis. The generated AP and ion current together with corresponding derivatives calculated for a dimensionless FHN equation are shown in Fig. 9.1. The assumption on using derivative J_T as a driving force has an important property – the force exerted to the biomembrane is bipolar and is therefore energetically balanced. If a localised pulse-type driving force is used then the energetical balance due to the moving signal $Z(X, T)$ is distorted by the continuous energy influx.

The AP coupled to the LW only

In this case, we neglect the PW in the axoplasm and formulate a model including the AP in a fibre and the accompanying LW in a biomembrane. Then the coupled model

Fig. 9.1 The solutions and their derivatives of the FHN equation. Top panel – action potential Z, recovery current J and their gradients Z_X, J_X in space at $T = 800$, bottom panel – action potential Z, recovery current J and their time derivatives Z_T, J_T in time at spatial node $n = 512$. Reproduced with permission from [5]; ©Springer Nature 2018.

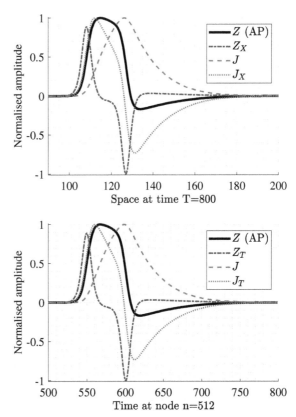

includes Eqs. (9.2), (9.4) and (9.7). The force $F_3(Z, J, P)$ is taken as $F_3(Z, J)$, i.e., depending only on the AP.

The main features are the following:

– the input (initial condition) for Eqs. (9.2) in the form of a narrow sech^2-type pulse with an amplitude above the threshold triggers the AP like all-or-nothing [15, 24];

– the generated electrical pulse (AP) has a typical asymmetric form with an overshoot and generates an ion current;

– the gradient (i.e., the change) of the ion current is taken as an input for the generation of the mechanical longitudinal wave (LW);

– the derivative of the LW gives the profile of the TW [6].

Note that (i) the gradient of the ion current is energetically balanced; (ii) the velocities of the AP and LW are chosen to be synchronised.

The simulation results in the dimensionless form are shown in Fig. 9.2 which demonstrates the profiles of the AP, LW and TW together with the ion current. The latter has a characteristic shape measured in several studies [11, 19, 29].

Fig. 9.2 Action potential coupled with the mechanical wave only ($\eta_1 = \eta_2 = 0$). Solutions at $T = 1500$. The coupling parameters are $\beta_1 = \beta_2 = 0.05$, $\gamma_1 = 0$, $\gamma_2 = 0.002$. Reproduced with permission from [5]; ©Springer Nature 2018.

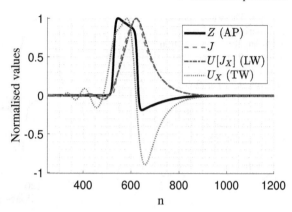

The AP coupled to the PW only

In this case, we formulate a model in terms of the electrical signal AP and the pressure wave PW. The model involves Eqs. (9.2), (9.3) and (9.6).

The simulation results are shown in Fig. 9.3 and the pressure profiles for different combinations of coupling parameters η_1 and η_2 are shown in Fig. 9.4. The pressure wave (PW) modelled by Eq. (9.3) demonstrates retardation from the AP and a slight overshoot [31]. As far as wave equation (9.3) has pretty stable solutions, the small changes in the coefficients η_1, η_2 which characterise the driving force F_2, do not lead to essential changes in the profile of the PW (see Fig. 9.4). Increasing η_1 leads to a steeper front and faster decay at the back of the profile, while the effect of the η_2 is opposite.

Fig. 9.3 Action potential coupled with the pressure wave (two different coupling forces considered). Solutions at $T = 1500$. The coupling parameters are $\beta_1 = \beta_2 = 0$, $\gamma_1 = \gamma_2 = 0$, $\eta = 0.002(Z_X)$, $\eta = 0.02(J_X)$. In the case of $P[Z_X, J_X]$ the coupling parameters are $\eta_1 = 0.001(Z_X)$, $\eta_2 = 0.01(J_X)$. Reproduced with permission from [5]; ©Springer Nature 2018.

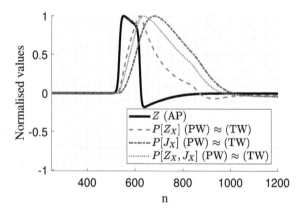

Fig. 9.4 Pressure wave profiles with different coupling parameters at $T = 1500$. Reproduced with permission from [5]; ©Springer Nature 2018.

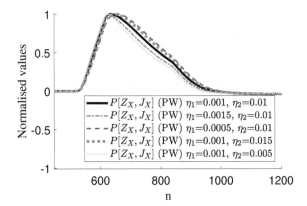

9.3 The AP Coupled to All Mechanical Effects

In this case, three components of a signal – AP, PW, LW – are taken into account [5, 8]. An important question is to estimate the forms of physically plausible coupling forces F_2 and F_3. From the analysis of the case with coupled AP and LW, it is possible to conclude that the character of the F_3 should be bipolar. The numerical simulation permits to calculate the profiles with several forces depending on Z_X, P_T, J_T. The corresponding wave profiles at $T = 1500$ are shown in Fig. 9.5 for the case when time derivatives are used for coupling forces and in Fig. 9.6 for the cases where mostly gradients are used as coupling terms.

In Fig. 9.6 the following phenomena are demonstrated:

(a) The pressure wave is generated by the action potential gradient and the mechanical wave is generated by the pressure time derivative. The coupling parameters are $\beta_1 = \beta_2 = 0.05$, $\gamma_1 = 0.002$, $\gamma_2 = 0$, $\eta_1 = 0.002$, $\eta_2 = 0$.

(b) The pressure wave is generated by the action potential gradient and the mechanical wave is generated by the pressure time derivative and the ion current gradient. The coupling parameters are $\beta_1 = \beta_2 = 0.05$, $\gamma_1 = \gamma_2 = 0.002$, $\eta_1 = 0.002$, $\eta_2 = 0$.

(c) The pressure wave is generated by the ion current gradient and the mechanical wave is generated by the pressure time derivative. The coupling parameters are $\beta_1 = \beta_2 = 0.05$, $\gamma_1 = 0.002$, $\gamma_2 = 0$, $\eta_1 = 0$, $\eta_2 = 0.02$.

(d) The pressure wave is generated by the ion current gradient and the mechanical wave is generated by the pressure time derivative and ion current gradient. The coupling parameters are $\beta_1 = \beta_2 = 0.05$, $\gamma_1 = \gamma_2 = 0.002$, $\eta_1 = 0$, $\eta_2 = 0.02$.

(e) The pressure wave is generated by the ion current gradient and action potential gradient while the mechanical wave is generated by the pressure time derivative and ion current gradient. The coupling parameters are $\beta_1 = \beta_2 = 0.05$, $\gamma_1 = \gamma_2 = 0.002$, $\eta_1 = 0.001$, $\eta_2 = 0.01$.

After the ensemble shown in Fig. 9.6 has emerged (initial phase not shown), it keeps together for a long run because the coupling forces are balanced. Mathematically the model supports also solutions where the waves in the ensemble have

Fig. 9.5 The solutions of the three wave model at $T = 1500$ when using the ion current time derivative J_T (top panel) and J_T plus pressure time derivative P_T and action potential gradient Z_X as a coupling forces (bottom panel). Parameters $\beta_1 = \beta_2 = 0.05$, $\gamma_1 = 0$, $\gamma_2 = 0.01$, $\eta_1 = 0$, $\eta_2 = 0.01$ (top panel) and $\beta_1 = \beta_2 = 0.05$, $\gamma_1 = 0.001$, $\gamma_2 = 0.01$, $\eta_1 = 0.001$, $\eta_2 = 0.01$ (bottom panel). Reproduced with permission from [5]; ©Springer Nature 2018.

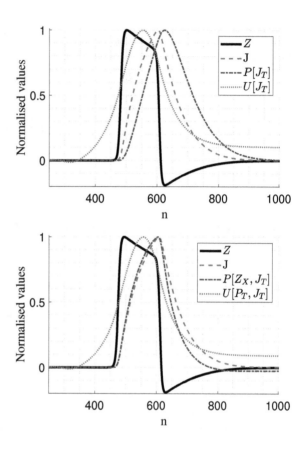

different limiting velocities including the cases where either the PW or the LW propagates faster than the AP.

From the viewpoint of the behaviour of the solution, there is almost no qualitative difference if we use a time derivative or spatial gradient as the coupling term because as demonstrated in Fig. 9.1, the shape of the function in essence is the same. In the used numerical scheme the calculation of spatial derivatives is more convenient and numerically more accurate and for that reason, in the following analysis, the focus is on the case where J_X is used as one of the coupling terms. Numerically we find J_X by making use of the properties of the fast Fourier transform while for finding J_T a simple backward difference scheme is used. However, if the experiments demonstrate the need to use J_T in coupling forces, this can also be realised.

The profiles in Figs. 9.5 and 9.6 demonstrate a typical AP with an overshoot, the PW propagating behind the AP and the LW in the biomembrane with a typical solitary wave profile. Feedback coupling is taken into account for the AP from the LW and its influence is more evident in Figs. 9.6b and 9.6e. These profiles correspond qualitatively to previous studies starting from the AP [16, 24] to experimentally

Fig. 9.6 The solutions of the three wave model when using the ion current gradient J_X as one of the coupling forces (cases a to d). See text for parameter details. Reproduced with permission from [5]; ©Springer Nature 2018.

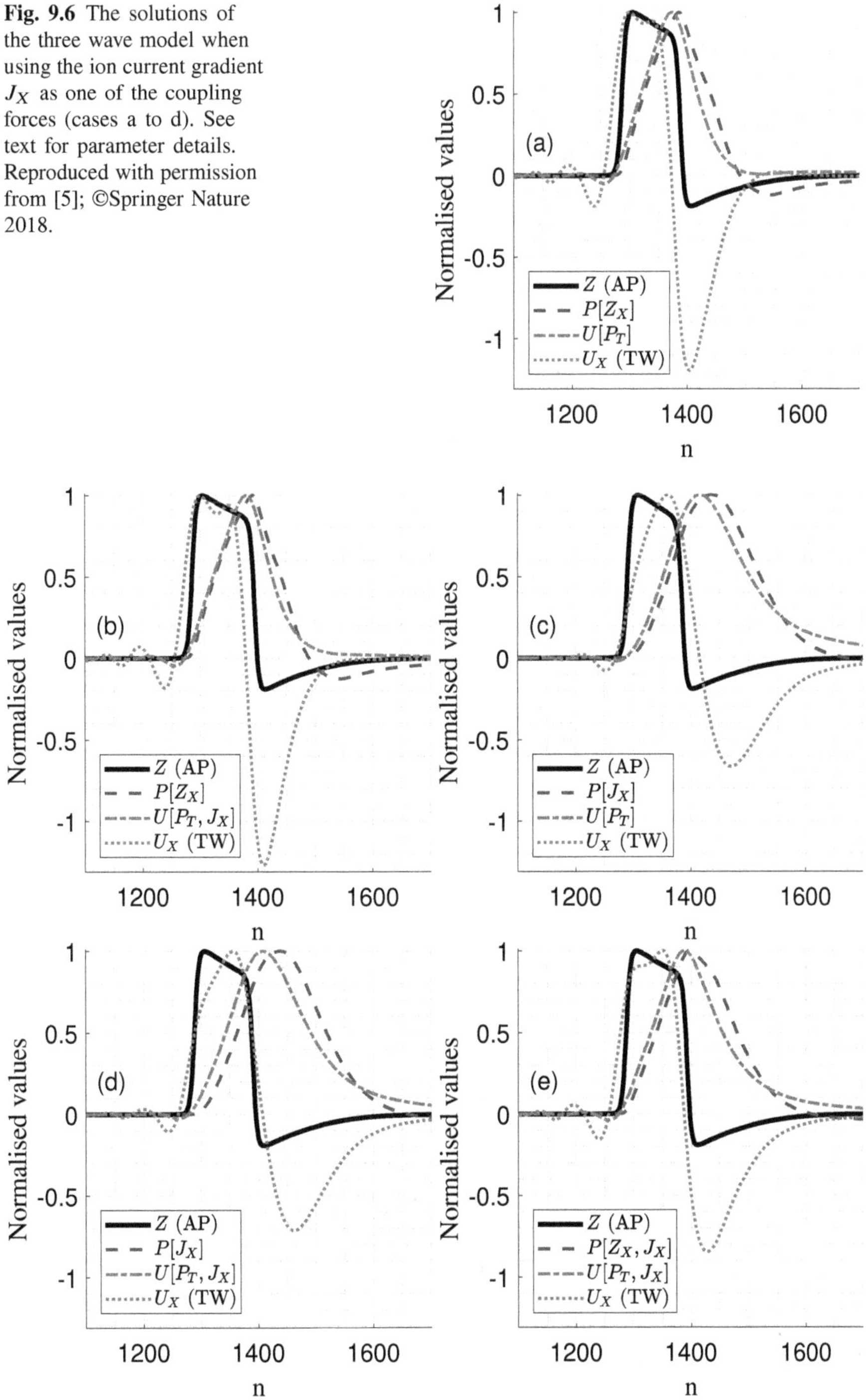

measured PW [31] and LW [11, 13]. The transverse wave TW is calculated from the LW by using $W \propto U_X$ (Eq. (9.5)) and has a bipolar shape [29]. The slight delay of the PW [31] is caused by chosen velocities but in principle could be adjusted according to measurements. Following the original Heimburg-Jackson model [13] a pretty wide LW has been calculated, but the improved model [7] with the dispersion term accounting for the inertia of lipids leads to a narrower pulse. In our computations, it is comparable with the width of the AP. Note that all the profiles are normalised with their maximal amplitude taken as a scaling measure.

Without the coupling forces, the AP, PW and LW are stable as expected. The numerical solutions of the ensemble are stable for a long period of simulations. However, like in the real system, the ensemble demonstrates slight variations but in terms of nonlinear dynamics [25] keeps asymptotic stability. It is interesting to note that as far as the equation for the LW describes also the emergence of soliton trains (splitting a pulse into a sequence of narrow solitons), the calculations for a very long time demonstrate the start of such an emergence [7].

The basic assumption in all calculations is that the coupling is influenced by the changes of the field quantities, not by their values. This idea is supported by several studies [11, 21, 22, 31]. The initial stage of the AP forming from an input is not analysed because of possible fast changes and presented analysis takes a fully formed AP as a basic signal for coupled waves. The profiles in Figs. 9.5-9.6 are qualitatively similar to all measured ones. The parameters for simulation shown in Figs. 9.5-9.6 have been chosen to generate mechanical effects a little bit behind the AP. This brings up the question about the synchronisation of velocities. In principle, the wave velocities in continua depend on elastic properties and density but due to coupling effects the group velocities (responsible for energy propagation) may considerably differ from the sound velocity due to dispersion. The velocity of the AP may also be affected by axonal irregularities and ion channels [4]. Note also that the velocity of the blood flow in a vessel depends on the stiffness of the vessel wall [2]. So, we have to agree that "the conduction velocity of mechanical impulses in nerve fibres is unknown" [2] and needs further theoretical and experimental studies in order to establish joint understanding. So, as a proof of concept, this demonstrates the idea to look for the biomembrane mediated signalling in a nerve fibre as a complex system, resulting in an ensemble of waves.

It can be concluded from profiles shown in Figs. 9.5-9.6 that the influence of changes for coupling can be presented either by gradients Z_X, J_X or by the time derivatives P_T, J_T or in the more general case as some combination of the considered coupling terms. In principle, it is possible to include some scale factors into the coupling forces. The reasoning is based on the assumption that if the biomembrane is deformed then moving it back to equilibrium might be easier due to elasticity of it (like in case of a nonlinear spring). As far as the simulations showed no significant effect of such modifications, these results are not demonstrated here. Another interesting question is about the strength of coupling forces. Here, the coupling parameters are chosen to be at least one order of magnitude smaller than other parameters in the governing equations: the mechanical activation coefficients are roughly 5% and the force coupling parameters roughly 1% compared to the largest parameters in governing equations. If the coupling parameters significantly above these values are used then the loss of stability was recorded in simulations. One of the reasons of such behaviour is related to high values of derivatives in the formation period of the ensemble from the spark-like input for emerging a stable AP. The estimates for the coupling parameters may also give a clue for experimental verification.

Note that the AP (that is Z) and ion currents (here only J) were measured already by Hodgkin [15] and the corresponding measurements reported in several recent overviews [3, 4]. This is related to functional dependencies. However, the coefficients of coupling forces F_1 and F_2 need calibrating and should be determined by the integral contribution of the effects. This means answering the question of whether the generated waves reach their measured values.

9.4 Temperature Changes Accompanying the AP

The temperature Θ is governed by the classical heat equation (see Sect. 8.1):

$$\Theta_T = \alpha \Theta_{XX} + F_4(Z, J, U, P). \tag{9.8}$$

The coupling forces are given by Eqs. (8.8)-(8.11) where the concept of internal variables [9, 10] is used for describing endothermic and exothermic processes. These equations and expressions represent our best understanding of the mechanism behind heat production [9, 10].

In this Section, however, we turn our attention to all mathematically possible cases. For this purpose, the following functions F_4 are used [28]:
(i) $F_4 = \tau_1 Z$ and $F_4 = \tau_2 Z^2$ (see Fig. 9.7);
(ii) $F_4 = \tau_3 J$ and $F_4 = \tau_4 J^2$ (see Fig. 9.8);
(iii) $F_4 = \tau_5 U$ and $F_4 = \tau_6 U^2$ (see Fig. 9.9);
(iv) $F_4 = \tau_7 Z_T + \tau_8 J_T$ and $F_4 = \tau_9 J_T + \tau_{10} U_X$ (see Fig. 9.10),
where τ_i are the thermal coupling coefficients.

For calculating temperature changes, some more remarks are in order. Note that in Figs. 9.7-9.10 thermal energy change Q is plotted as $Q = \Theta_X$. One can observe in Fig. 9.7 that in the case of temperature increase proportional to Z the local temperature drops in the negative polarity region of the AP while if the temperature increases proportional to Z^2, the local temperature keeps increasing even in the negative polarity region of the AP. Observing correlation between Z (or Z^2) and the measured temperature increase near axon seems to be common for several experiments by Howarth et al. [18], Ritchie et al. [26] and Tasaki and Byrne [30]. Experimental results published so far seem inconclusive to argue firmly in favour of either option as normally only some kind of averaged thermal energy production over some time can be measured. Another note to be made is that while experimentally the heat production proportional to AP has been observed, this correlation does not have to mean direct causality. Like generally in nature, the potential gradient alone is

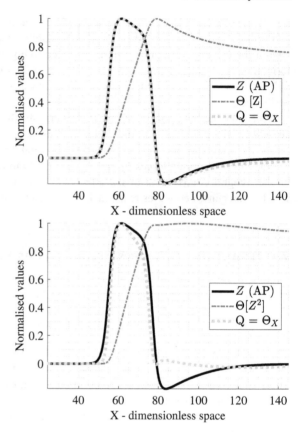

Fig. 9.7 The temperature change Θ and thermal energy change $Q = \Theta_X$ with coupling force $F_4 = \tau_1 Z$ (top panel) and $F_4 = \tau_2 Z^2$ (bottom panel). Note that profiles of Z and Q partly coincide. Reproduced with permission from [28]; ©Walter de Gruyter and Company 2019.

rarely the only source of the temperature change in an environment. One could argue that heat production might instead be proportional to ion currents (Fig. 9.8) or the longitudinal density change (Fig. 9.9) which accompany the propagating AP. For the sake of completeness temperature increase as a function of J and J^2 is presented in Fig. 9.8 and as a function of density change U and U^2 in Fig. 9.9. The idea behind including J and U as sources is that mechanisms, where the temperature increase is caused by the current flowing through the environment or by the deformation of the solid, are well established in the physics even if these are not as common sources assumed for the heat generation [14, 23] as the correlation with the AP [18, 26, 30].

In such a context it might not be a problem to observe that some heat energy 'goes away' in the negative polarity phase of the used driving signal if one assumes heat production proportional to the driving signal as the underlying endothermic mechanism might be independent of the AP propagation, like, some kind of endothermic chemical reaction [1]. The open question in the latter case is, however, the issue of time scales. Moreover, another effect present in the numerical simulation results presented must be noted – at least in part the reason why the temperature can be lower in the areas through which the signal has already propagated as opposed to the

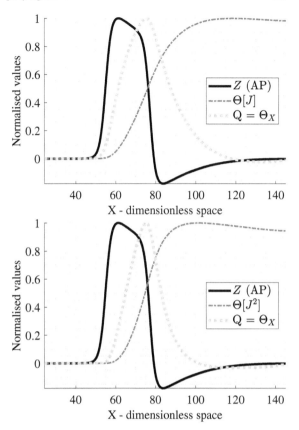

Fig. 9.8 The temperature change Θ and thermal energy change $Q = \Theta_X$ with coupling force $F_4 = \tau_3 J$ (top panel) and $F_4 = \tau_4 J^2$ (bottom panel). Reproduced with permission from [28]; ©Walter de Gruyter and Company 2019.

front is that the driving signal can be actually of lower amplitude at earlier stages of the signal propagation. This is the main reason why in Fig. 9.9 there is quite a significant drop in temperature for the case of $F_4 \propto U^2$ (bottom panel) because of the changes in driving signal amplitude. The signal shape in Fig. 9.9 (bottom panel) is qualitatively similar to the signal shapes observed by Abbott et al. [1]. Another note relevant specifically for the Eq. (9.4) can be made – it is a conservative equation before adding the coupling term F_3 which facilitates energy exchange between the coupled equations. However, if the mechanical wave in the lipid bi-layer is considered as a source for thermal energy in the system for earnest this equation would need some kind of additional term which would take some energy away from the mechanical wave. For example, the simplest possibility might be using a Voigt–type model which means including a term U_{XXT} in Eq. (9.4).

Experimentally the thermal energy production and consumption has been observed in the time scales comparable to the AP propagation [29, 30] up to few orders of magnitude longer as shown by Howarth et al. [18]. Following the idea that the underlying processes responsible for the heat production and consumption might be proportional to some of the processes we are actually modelling one possibility

Fig. 9.9 The temperature
change Θ and thermal energy change $Q = \Theta_X$ with
coupling force $F_4 = \tau_5 U$
(top panel) and $F_4 = \tau_6 U^2$
(bottom panel). Reproduced
with permission from [28];
©Walter de Gruyter and Company 2019.

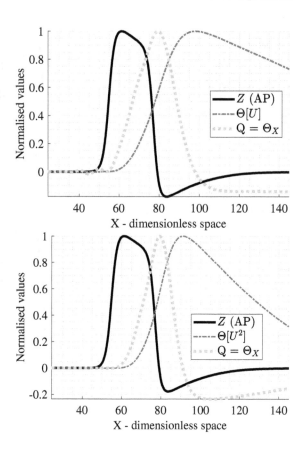

shown in Fig. 9.10 where time derivatives of AP and ion current J (top) or the ion
current time derivative and the gradient of longitudinal density change (bottom) are
used for the thermal energy generation and consumption. In such a case thermal energy is generated when the driving signal is growing and consumed when the driving
signal is decreasing. The examples shown in Fig. 9.10 are intended to highlight what
is mathematically possible in the present framework and not as an implication of the
thermal response being actually generated and even more importantly, consumed,
by the used signals in the reality. However, some of the experimental observations
are qualitatively similar to profiles shown in Fig. 9.10 [30].

9.5 Full Coupled Model

The full coupled model (8.1)-(8.11) for signal propagation in nerve axons is presented
in Sect. 8.1. The AP is described by the FHN equation (8.1), PW by a modified wave
equation (8.2), LW by the improved HJ model (8.3) and temperature Θ by the

Fig. 9.10 The temperature change Θ and thermal energy change $Q = \Theta_X$ with coupling force $F_4 = \tau_7 Z_T + \tau_8 J_T$ (top panel) and $F_4 = \tau_9 J_T + \tau_{10} U_X$ (bottom panel). Reproduced with permission from [28]; ©Walter de Gruyter and Company 2019.

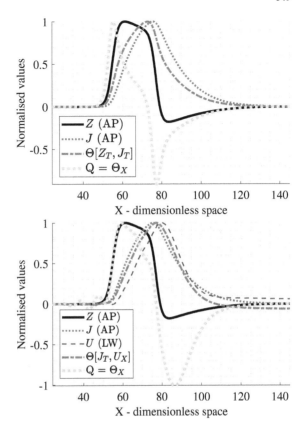

classical heat equation (8.5). Transverse displacement TW is calculated like in the theory of rods $W \propto U_X$ (Eq. (8.4)). In addition there are coupling forces F_2, F_3 and F_4 given by Eqs. (8.6)-(8.8). The endo- and exothermic effects in coupling force (8.8) are accounted through Eqs. (8.9)-(8.11) by making use of the concept of internal variables (see Sect. 7.5).

An example of the cases where only a single source of thermal energy is present is shown in Fig. 9.11. The profiles for the Z, P, U are the same but which is different, is the thermal response Θ curves. Qualitatively what is relevant here is the location of the thermal response peak compared to the wave ensemble and the shape of the initial thermal response curve.

Two distinct thermal response cases are shown in Fig. 9.12. It must be noted that the PW and LW profiles are also significantly different. A case with full F_4 in the form of $\tau_1 J^2 + \tau_2 (P_T + \varphi_2(P)) + \tau_3 (U_T + \varphi_3(U)) - \tau_4 \Omega$ and the viscosity in P (8.2) and U (8.3) is shown in top panel in Fig. 9.12. In the bottom panel in Fig. 9.12 the irreversible thermal processes have been significantly suppressed by setting coefficient τ_1 to zero (related to the Joule heating) and reducing the viscosity by two orders of magnitude compared to the case in the top panel. This means that the thermal response in Fig. 9.12 – see the bottom panel – is dominated by the

Fig. 9.11 Wave ensemble Z, P, U, Θ with different temperature models. Top panel depicts the case $\Theta \propto J^2$, where $\tau_1 = 0.05$, $\tau_2 = 0$, $\tau_3 = 0$; middle panel – the case $\Theta \propto \varphi_2(P)$, where $\tau_1 = 0$, $\tau_2 = 0.1$, $\tau_3 = 0$; bottom panel – the case $\Theta \propto \varphi_3(U)$, where $\tau_1 = 0$, $\tau_2 = 0$, $\tau_3 = 0.5$. Reproduced with permission from [10]; ©Estonian Academy of Sciences 2020.

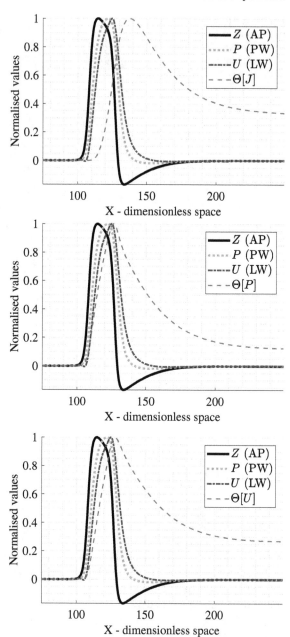

Fig. 9.12 Wave ensemble Z, P, U, Θ with different thermal sources. Top panel depicts the case of small, but noticeable viscous dampening in P and U; thermal sources in F_4 are J^2, $\varphi_2(P)$, $\varphi_3(U)$, where $\tau_1 = 0.0167$, $\tau_2 = 0.0333$, $\tau_3 = 0.1667$. Bottom panel depicts the case where the thermal sources in F_4 are only from P_T and U_T with negligible viscosity. Here $\tau_1 = 0$, $\tau_2 = 0.1$, $\tau_3 = 0.025$, $\mu_1 = 0.0001$, $\mu_2 = 0.0001$. Reproduced with permission from [10]; ©Estonian Academy of Sciences 2020.

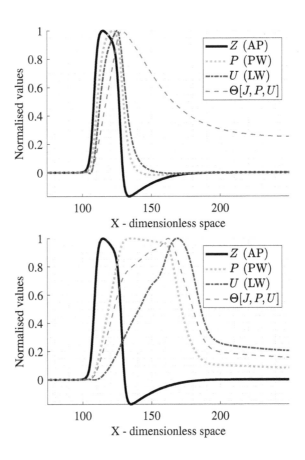

reversible processes (temperature increase when density increases in axoplasm and membrane and temperature decrease if the density decreases with friction related terms suppressed). It can also be noted that viscosity or lack of it has a significant effect on the shape of the PW and LW profiles within the range of the used parameters. The governing equations for the PW and LW are conservative before adding the coupling forces F_2, F_3 and the dissipative terms. Without accounting for some kind of friction-like term in PW and LW the different sound velocities in the corresponding environments have an opportunity to spread the wave-profiles out as the driving force from the AP is taken with a different velocity than the sound speed in axoplasm and biomembrane. While not demonstrated in the figures, it should be noted the thermal signal drop off rate is dependent on the relaxation time τ_Ω. If relaxation is taken fast enough it is possible to model thermal response curves which are almost perfectly in-phase with the propagating wave ensemble. Some endothermic chemical processes in the context of the nerve pulse propagation have been discussed by Abbott et al. [1].

Fig. 9.13 The effect of Joule heating ($F_4 \propto Z^2$). Top panel depicts the wave profiles Z, P, U, Θ. Here for $\tau_1 Z^2 \approx \Omega$ the parameters are $\tau_1 = 0.0005$, $\tau_2 = 0$, $\tau_3 = 0$. Middle panel depicts how residual heat depends on the balance between thermal source and thermal sink; for $\tau_1 Z^2 > \Omega$ the parameters are $\tau_1 = 0.001$, $\tau_2 = 0$, $\tau_3 = 0$; bottom panel – the effect of relaxation time on the temperature profiles; here the parameters are chosen so that $\tau_1 Z^2 \approx \Omega$. Reproduced with permission from [10]; ©Estonian Academy of Sciences 2020.

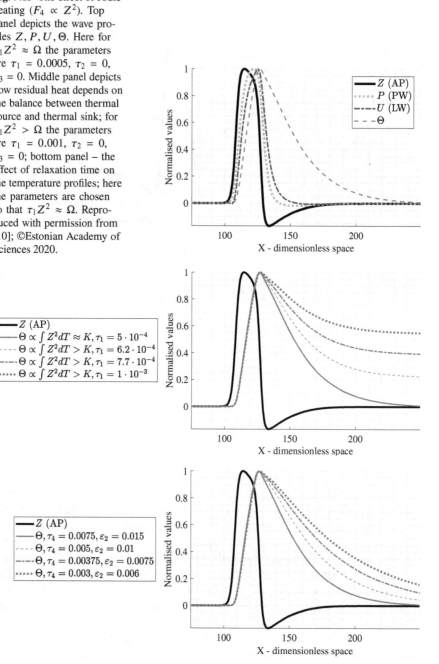

Alternatively, thermal response in-phase with the propagating wave ensemble can emerge if the reversible thermal changes are dominating (Fig. 9.12 bottom panel).

Figure 9.13 demonstrates the effect of the Joule heating as a function of Z^2. In the top panel, the thermal source and thermal sink have been balanced $\tau_1 Z^2 \approx \Omega$ so that the thermal response settles eventually back to its starting value. In the middle panel, the balance between thermal source and thermal sink terms is varied, demonstrating that depending on the balance between these terms there is a different residual level of heat increase after the nerve pulse wave ensemble has passed. In the bottom panel, the thermal source and thermal sink are balanced and the relaxation time is varied. It demonstrates that depending on the relaxation time related parameters, the equilibrium or residual level can be reached at different rates. In terms of qualitative thermal response curve characteristics, there is no practical difference between formulating the Joule heating in terms of Z^2 or J^2.

9.6 Dimensionalised Example

Fig. 9.14 Dimensionalised AP with $z_{ap} = 100$ [mV] and $t_{ap} = 0.05$ [ms].

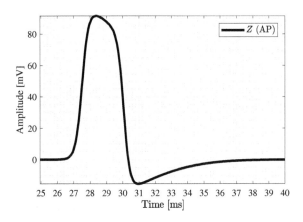

It is clear that the solution of dimensionless governing equations in simulations permits only qualitative estimations. In other words, such comparison serves as a proof of concept. In order to compare the simulation results directly to experimental studies, the governing equations must be written in their dimensional form (see Chap. 8). Such an approach faces two kinds of difficulties. The first problem is related to the great number of coefficients. A good example is the HH model, which includes coefficients characterising conductances and relaxation coefficients of phenomenological variables [17, 20]. If a certain set of these parameters is chosen then it means choosing a concrete axon and the other components of the wave ensemble might not be experimentally studied. The second problem is related to the complexity of the wave ensemble which needs more experiments [12], for example, for characterising the contact forces with their physical coefficients. As shown above, in this Chapter, the modelling *in silico* permits to carry on numerical experiments at the wide scale of parameters. Consequently, the estimation of the full set of needed parameters needs more experimental studies. However, here we present an approach to 'dimensionalise' the results obtained so far by numerical simulation.

For example, let us assume that we have the following experimental observations: (i) The amplitude of the AP has been measured as 100 [mV], the duration of the AP main pulse has been measured to be 5 [ms] and the velocity of the AP has been measured to be 50 [m/s], (ii) the pressure change in axoplasm has been measured to be 1 [mPa], (iii) the transverse displacement of the axon membrane has been measured to be 1 [nm], (iv) the temperature change accompanying the single nerve signal ensemble has been assessed to be 15 [μK]. Knowing these quantities allows straightforward 'scaling' of the dimensionless results by picking a suitable coefficient (with the corresponding dimension) to multiply the dimensionless result with. However, one should note that not all parameters are independent between the members of the ensemble, as, for example, space and time should be of the same scale for all modelled quantities.

The same procedure can be repeated also for other signals in the ensemble by picking the scaling coefficents with dimensions so that the dimensionless results scale to the desired values (see Figs. 9.15 and 9.16).

Fig. 9.15 Dimensionalised LW with $u_{\mathrm{lw}} = 10$ [ ] and TW with $w_{\mathrm{tw}} = 1000$ [nm].

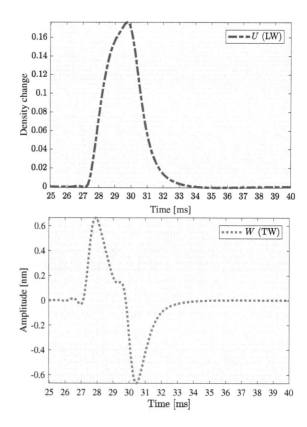

Fig. 9.16 Dimensionalised
PW with $p_{pw} = 15$ [mPa] and
Θ with $\theta_{tmp} = 1000$ [μK].

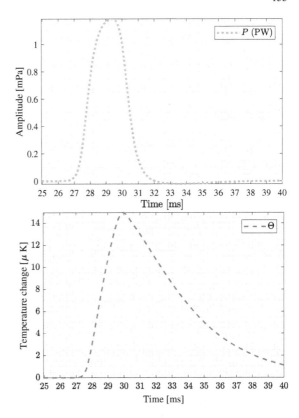

Let us say, that observing the dimensionless simulation results in time at a spatial
node $n = 512$ (while the initial 'spark' was given centred at node $n = 1024$) one notes
that the dimensionless amplitude of the Z is 0.9151, dimensionless duration of the
AP main pulse is 200 time units. For getting an AP with the desired characteristics
one can then pick coefficients as $z_{ap} = 100$ [mV] and $t_{ap} = 0.05$ [ms], multiply Z
and T by these coefficients resulting in an example given on Fig. 9.14. Knowing the
velocity of the signal would allow one to also calculate the spatial scaling if needed,
as one can note that it took about 27 ms for the AP pulse to form from the initial
condition and to propagate from node $n = 1024$ (initial condition centre point) to a
node $n = 512$.

The parameters for the dimensionless numerical simulation which is dimensionalised in Figs. 9.14 - 9.16 are the following: $n = 2048$, $L = 128$, $\Delta T = 1$, $D = 1$,
$\epsilon = 0.018$, $a_1 = a_2 = 0.2$, $\beta_1 = \beta_2 = 0.025$, $c_3^2 = 0.1$, $N = -0.05$, $M = 0.02$,
$H_1 = 0.2$, $H_2 = 0.99$, $\mu_3 = 0.05$, $\gamma_1 = 0.01$, $\gamma_2 = 10^{-3}$, $\gamma_3 = 10^{-5}$, $c_2^2 = 0.09$,
$\mu_2 = 0.05$, $\eta_1 = 10^{-3}$, $\eta_2 = 0.01$, $\eta_3 = 10^{-3}$, $\alpha = 0.005$, $\tau_1 = 5 \cdot 10^{-4}$, $\tau_2 = \tau_3 = 0$.
The initial condition used here was the standard sech^2 pulse of Z in the middle of
spatial period with amplitude above threshold and all the other initial conditions
were set to zero (see Sect. 9.1).

The authors hope that the *in silico* simulations presented above will help the further experimental studies. The results shown in Figs. 9.14, 9.15 and 9.16 demonstrate a good match with experimentally measured profiles in Chap. 4.

References

1. Abbott, B.C., Hill, A.V., Howarth, J.V.: The positive and negative heat production associated with a nerve impulse. Proc. R. Soc. B Biol. Sci. **148**(931), 149–187 (1958). DOI 10.1098/rspb.1958.0012
2. Barz, H., Schreiber, A., Barz, U.: Impulses and pressure waves cause excitement and conduction in the nervous system. Med. Hypotheses **81**(5), 768–72 (2013). DOI 10.1016/j.mehy.2013.07.049
3. Clay, J.R.: Axonal excitability revisited. Prog. Biophys. Mol. Biol. **88**(1), 59–90 (2005). DOI 10.1016/j.pbiomolbio.2003.12.004
4. Debanne, D., Campanac, E., Bialowas, A., Carlier, E., Alcaraz, G.: Axon physiology. Physiol. Rev. **91**(2), 555–602 (2011). DOI 10.1152/physrev.00048.2009.
5. Engelbrecht, J., Peets, T., Tamm, K.: Electromechanical coupling of waves in nerve fibres. Biomech. Model. Mechanobiol. **17**(6), 1771–1783 (2018). DOI 10.1007/s10237-018-1055-2
6. Engelbrecht, J., Tamm, K., Peets, T.: On mathematical modelling of solitary pulses in cylindrical biomembranes. Biomech. Model. Mechanobiol. **14**, 159–167 (2015). DOI 10.1007/s10237-014-0596-2
7. Engelbrecht, J., Tamm, K., Peets, T.: On solutions of a Boussinesq-type equation with displacement-dependent nonlinearities: the case of biomembranes. Philos. Mag. **97**(12), 967–987 (2017). DOI 10.1080/14786435.2017.1283070
8. Engelbrecht, J., Tamm, K., Peets, T.: Modeling of complex signals in nerve fibers. Med. Hypotheses **120**, 90–95 (2018). DOI 10.1016/j.mehy.2018.08.021
9. Engelbrecht, J., Tamm, K., Peets, T.: Internal variables used for describing the signal propagation in axons. Contin. Mech. Thermodyn. (2020). DOI 10.1007/s00161-020-00868-2
10. Engelbrecht, J., Tamm, K., Peets, T.: On mechanisms of electromechanophysiological interactions between the components of signals in axons. Proc. Estonian Acad. Sci. **69**(2), 81–96 (2020). DOI 10.3176/proc.2020.2.03
11. Gonzalez-Perez, A., Mosgaard, L., Budvytyte, R., Villagran-Vargas, E., Jackson, A., Heimburg, T.: Solitary electromechanical pulses in lobster neurons. Biophys. Chem. **216**, 51–59 (2016). DOI 10.1016/j.bpc.2016.06.005
12. Heimburg, T.: The important consequences of the reversible heat production in nerves and the adiabaticity of the action potential. arXiv:2002.06031v1 [physics.bio-ph] (2020)
13. Heimburg, T., Jackson, A.D.: On soliton propagation in biomembranes and nerves. Proc. Natl. Acad. Sci. USA **102**(28), 9790–5 (2005). DOI 10.1073/pnas.0503823102
14. Heimburg, T., Jackson, A.D.: Thermodynamics of the nervous impulse. In: N. Kaushik (ed.) Structure and Dynamics of Membranous Interfaces, chap. 12, pp. 318–337. John Wiley & Sons (2008)
15. Hodgkin, A.L.: The Conduction of the Nervous Impulse. Liverpool University Press (1964)
16. Hodgkin, A.L., Huxley, A.F.: Resting and action potentials in single nerve fibres. J. Physiol. **104**, 176–195 (1945)
17. Hodgkin, A.L., Huxley, A.F.: A quantitative description of membrane current and its application to conduction and excitation in nerve. J. Physiol. **117**(4), 500–544 (1952). DOI 10.1113/jphysiol.1952.sp004764
18. Howarth, J.V., Keynes, R.D., Ritchie, J.M.: The origin of the initial heat associated with a single impulse in mammalian non-myelinated nerve fibres. J. Physiol. **194**(3), 745–93 (1968). DOI 10.1113/jphysiol.1968.sp008434

19. Iwasa, K., Tasaki, I., Gibbons, R.: Swelling of nerve fibers associated with action potentials. Science **210**(4467), 338–339 (1980). DOI 10.1126/science.7423196

20. Izhikevich, E.M.: Dynamical Systems in Neuroscience. The MIT Press (2006). DOI 10.7551/mitpress/2526.001.0001

21. Kim, G.H., Kosterin, P., Obaid, A.L., Salzberg, B.M.: A mechanical spike accompanies the action potential in Mammalian nerve terminals. Biophys. J. **92**(9), 3122–9 (2007). DOI 10.1529/biophysj.106.103754

22. Mueller, J.K., Tyler, W.J.: A quantitative overview of biophysical forces impinging on neural function. Phys. Biol. **11**(5), 051001 (2014). DOI 10.1088/1478-3975/11/5/051001

23. Mussel, M., Schneider, M.F.: It sounds like an action potential: unification of electrical, chemical and mechanical aspects of acoustic pulses in lipids. arXiv:1806.08551 [physics.bio-ph] (2018)

24. Nagumo, J., Arimoto, S., Yoshizawa, S.: An active pulse transmission line simulating nerve axon. Proc. IRE **50**(10), 2061–2070 (1962). DOI 10.1109/JRPROC.1962.288235

25. Nicolis, G., Nicolis, C.: Foundations of Complex Systems. World Scientific, New Jersey et al. (2007)

26. Ritchie, J.M., Keynes, R.D.: The production and absorption of heat associated with electrical activity in nerve and electric organ. Q. Rev. Biophys. **18**(04), 451 (1985). DOI 10.1017/S0033583500005382

27. Salupere, A.: The pseudospectral method and discrete spectral analysis. In: E. Quak, T. Soomere (eds.) Applied Wave Mathematics, pp. 301–334. Springer Berlin Heidelberg, Berlin (2009). DOI 10.1007/978-3-642-00585-5

28. Tamm, K., Engelbrecht, J., Peets, T.: Temperature changes accompanying signal propagation in axons. J. Non-Equilibrium Thermodyn. **44**(3), 277–284 (2019). DOI 10.1515/jnet-2019-0012

29. Tasaki, I.: A macromolecular approach to excitation phenomena: mechanical and thermal changes in nerve during excitation. Physiol. Chem. Phys. Med. NMR **20**, 251–268 (1988)

30. Tasaki, I., Byrne, P.M.: Heat production associated with a propagated impulse in bullfrog myelinated nerve fibers. Jpn. J. Physiol. **42**(5), 805–813 (1992). DOI 10.2170/jjphysiol.42.805

31. Terakawa, S.: Potential-dependent variations of the intracellular pressure in the intracellularly perfused squid giant axon. J. Physiol. **369**(1), 229–248 (1985). DOI 10.1113/jphysiol.1985.sp015898

Chapter 10
Final Remarks

> *For a physicist mathematics is not just a tool by means of which phenomena can be calculated, it is the main source of concepts and principles by means of which new theories can be created.*
>
> *Freeman Dyson, 1964*

Biophysical processes are extremely complex and are characterised by a mixture of physical, chemical, and thermal effects. In order to understand the phenomena, there is a real need to work at the interface of many scientific disciplines. In physics, theory and experiments are strongly supporting each other, the same could be said about chemistry. In biology, many understandings are supported by the experiments. It has been said [2] "We want theories that engage meaningfully with the myriad of experimental details of particular systems, yet still are derivable from principles that transcend these details". One could also say that collecting the details into a whole is the idea of complex systems. This is the clear reason why nowadays much attention is paid to systems biology or biological complexity [15, 17]. It is quite understandable that one should try to unite the perspectives from physics [5] with those of biology [14]. On the other hand, this means also the usage of mathematical tools [12]. It is even said that mathematics is biology's next microscope [4].

In this book we have set up the general ideas of the phenomenological modelling of the propagation of signals in nerves: starting from the general philosophy of modelling of complex biological processes, we proceed to the casting of physical effects into the mathematical language, then the detailed analysis of governing equations of single effects is given and finally the analysis of a coupled system is presented (see Chap. 1, Introduction).

Although we keep in mind the famous quotation from A.Einstein: "Everything should be made as simple as possible but not simpler", it is not an easy task. The need for many parameters must also be treated critically like mentioned by J. von Neumann [8].

In what follows, is a summary of modelling with highlights stressed.

Nerve pulse propagation is a physical process in time. Whatever the mathematical models are constructed, they should be derived from basic principles (conservation of momentum) and *include time as one of the independent variables*. This is why we started from the description of the corresponding equations of mathematical physics. Without any doubt, the situation in nerves is much more complicated compared with the classical cases of wave motion but the knowledge from the classical analysis could help the further modelling. The important conclusion is that the influence

© The Author(s), under exclusive license to Springer Nature Switzerland AG 2021
J. Engelbrecht et al., *Modelling of Complex Signals in Nerves*,
https://doi.org/10.1007/978-3-030-75039-8_10

of other effects in the process affecting a wave, can be modelled by the coupling
(contact) forces in the equations of motion. The possible diffusion processes have
also time-dependence and can be influenced by contact forces.

The next step is to systemise the effects in nerves and the possible dynamical
variables measured experimentally. As a result, the following effects are picked up:
the action potential AP with its amplitude Z, the ion current J (note that there are
more ion currents in the process); the longitudinal wave LW in the biomembrane
with an amplitude U and the corresponding transverse displacement TW with an
amplitude W; the pressure wave PW in the axoplasm with an amplitude P; the
temperature change Θ. Most of the governing equations for these effects are well
studied, only the recently proposed governing equation for LW [13] needs special
attention in order to specify possible types of solutions.

The Heimburg-Jackson (HJ) equation is improved by adding proper dispersion
term based on the analysis of continuum mechanics [9] and later called the improved
Heimburg-Jackson (iHJ) equation. It has been shown that by the iHJ equation the
following additional points have been described: *(i) inertia of lipid molecules taken
into account; (ii) velocity is bounded for all frequencies; (iii) properties of solitonic
solutions analysed.* The latter means that like the HJ equation, the iHJ possesses the
solitonic solutions but *narrower pulses* dues to the influence of inertia are possible; in
addition, it has been shown that there may exist two-soliton solutions, soliton trains,
negative solitons and periodic solutions. It has been shown that the interaction of
solitons is inelastic. These results may be of interest for cell mechanics in general.

The systematic analysis of experimental results (Chap. 7) reveals the possible
interactions between the variables and possible interaction types. Based on such
an analysis, it is possible to build up a joint mathematical model at the interface of
physiology, physics and mathematics. This model is based on (i) *the Hodgkin-Huxley
paradigm* which takes the triggering of an electrical signal as a starting point for the
process and (ii) *electro-mechano-physiological interactions* between the components
of nerve signals. Mathematically it means that the governing equations are united
into a joint system by contact forces and these are based on the following hypotheses
(Chap. 8):

(i) the mechanical waves and the temperature change in a nerve fibre are generated
due to *changes in electrical signals* (AP or ion currents);

(ii) if there is no clear evidence about the mechanisms of interaction then the *concept
of internal variables* is used.

These hypotheses give the basis for constructing the functional dependencies
of contact forces which should model the mechanisms of interactions discovered in
numerous experiments. Still, one should add some explanations. First, the changes of
variables mean mathematically either the *space or time derivatives* (like Z_X, Z_T, J_X,
...). Second, the pulse-type profiles of electrical signals mean that the derivatives
have a *bi-polar shape* which is energetically balanced. Third, the possibility to use
internal variables permits to model *exo- and endothermic processes*. Fourth, although
the triggering of the process is due to an electrical signal, the *possible feedback* from
accompanying effects to the electrical signal is also possible.

These hypotheses and the preliminary analysis of experiments permit to build up a coupled mathematical model involving the dynamical changes of the AP, the LW, the TW, the PW and temperature Θ (Chap. 8). This model includes *primary and secondary components*. The primary components are the AP, the LW and the PW which are characterised by corresponding velocities and their governing equations are derived from a basic wave equation. The secondary components are the TW and temperature Θ which have no physical velocities but can be derived from the primary components. Following the theory of rods, it is proposed that the LW and the TW are related: the amplitude W is proportional to the gradient of U. The basic model for temperature Θ is derived from the diffusion (parabolic) equation. The primary and secondary components together form *an ensemble of waves with synchronised velocities*. Such an effect is seen from the experiments and is also a consequence from the modelling. Note that this is not a usual coupling where velocities or waves are different. An ensemble is actually a signature of *the complexity* which is not surprising in the *systems biology*.

Although analytical solutions exist for several basic cases (Chap. 3), the complexity of biological processes as demonstrated above (Chap. 7) leads to the wide usage of computational methods. The *in silico* experiments based on the usage of the pseudospectral method (Chap. 9) cover a wide range of examples and demonstrate good qualitative correspondence to experimental results. The accuracy of numerical methods must always be analysed (see Appendix A). Such analysis accompanies many studies [3, 16].

The coupled system described in detail in this book is certainly a robust one. However, it is physically grounded and mathematically clear. The principle of causality is clearly followed; from an initial electrical excitation above the threshold, the AP and ion currents are generated which influence the emerging of accompanying effects. The model can be improved in many details because there are many physiological effects not taken into account such as the influence of anaesthetic drugs, the role of proteins, etc.

It must be stressed that the initial set of conservation laws in continuum mechanics as a basic starting point of the modelling is a physically consistent system. It is also well known that the modelling will involve beside conservation laws also the constitutive relations (relations between the variables) and these should satisfy certain conditions like determinism, equipresence, etc. [11]. The equipresence means that all dependent variables must be dependent on the same list of dependent variables unless the changes in their physical values are not significantly affected. In modelling described above the changes in the geometry of fibre due to the signal propagation are considered so small that at this stage they are neglected. However, the feedback ideas are described and if experimental studies will demonstrate the essential changes then it is possible to account for additional modifications.

We shall repeat a quotation from the Introduction (Chap. 1): there is a need *"to frame a theory that incorporates all observed phenomena in one coherent and predictive theory of nerve signal propagation"* [1]. What we presented is an attempt to build up such a theory at the interface of physiology, physics and mathematics, using the knowledge from neighbouring areas of studies describing the dynamical processes. The keywords which characterise the novel ideas are marked above in italics.

The approach described in this book can be generalised for modelling complex biological processes. where dynamics plays a leading role. Noble [15] indicates three levels of criteria in modelling: 1) descriptive; 2) integrative; 3) explanatory. In this context, the second level should answer the question – how do all the elements of a model interact. Based on nerve signal modelling, the following *basic principles of modelling complex biological processes* can be formulated:

- physics and the consistent conservation laws (continuum mechanics) form the basis for describing dynamical processes in biology resulting in wave-type and diffusion-type governing equations;

- the interactions between the fields (electrical, mechanical and thermal) are governed by changes in one field (generated by an external excitation) which trigger the changes in other fields;

- the physical changes are mathematically described by space and time derivatives of variables characterising the fields;

- internal variables might be useful for describing phenomena which are too complicated to be described directly but are characterised by their influence in changing the observable variable from one level to another;

- the mathematical system of governing equations describing a complex biological system includes the forces which describe the coupling effects.

Mathematical modelling is used to describe physical reality. From a general viewpoint, this description is actually a deep philosophical problem. Within the context of this book, possible keywords for the philosophical analysis are dynamics and complexity. Deleuze and Guattari [7] have analysed the ontology (related to the nature of being) and epistemology (concerned with knowledge) of dynamical processes. Their philosophy is reconstructed in terms of mathematics and nonlinear dynamics by DeLanda [6]. He stresses the notion of multiplicity characterised by differences which are productive and cause interaction. One should distinguish between intensive and extensive properties of systems; intensive properties like pressure, temperature, density, etc. cannot be divided, extensive properties like length, area, volume, amount of energy can be divided into parts. Non-equilibrium (intensive by DeLanda) states demonstrate explicitly the potential of nonlinearities which do not cause essential differences in equilibrium states or close to them. One should also distinguish between intrinsic (belonging to the system) and extrinsic (originating from outside) conditions for a system. These descriptions match well the process of modelling the propagation of signals in nerves described in this book. *The ensemble of waves is formed due to changes in intrinsic intensive variables of the system.* The ideas of this modelling in terms of DeLanda [6] are described by Engelbrecht et al. [10]. In the nutshell, intensive differences are productive [6] – this is actually the essence of the philosophy behind modelling of signals in nerves. Such a situation occurs actually in many dynamical biological processes and helps to understand the emergent properties [14]. The propagation of signals in nerves is an impressive example of interdisciplinarity and multiplicity.

References

1. Andersen, S.S.L., Jackson, A.D., Heimburg, T.: Towards a thermodynamic theory of nerve pulse propagation. Prog. Neurobiol. **88**(2), 104–13 (2009). DOI 10.1016/j.pneurobio.2009.03.002
2. Bialek, W.: Perspectives on theory at the interface of physics and biology. Reports Prog. Phys. **81**(1), 012601 (2018). DOI 10.1088/1361-6633/aa995b
3. Chen, H., Garcia-Gonzalez, D., Jérusalem, A.: Computational model of the mechanoelectro-physiological coupling in axons with application to neuromodulation. Phys. Rev. E **99**(3), 032406 (2019). DOI 10.1103/PhysRevE.99.032406
4. Cohen, J.E.: Mathematics is biology's next microscope, only better; biology is mathematics' next physics, only better. PLoS Biol. **2**(12), e439 (2004). DOI 10.1371/journal.pbio.0020439
5. Coveney, P.V., Fowler, P.W.: Modelling biological complexity: A physical scientist's perspective. J. R. Soc. Interface **2**(4), 267–280 (2005). DOI 10.1098/rsif.2005.0045
6. DeLanda, M.: Intensive Science and Virtual Philosophy. Continuum, London (2002).
7. Deleuze, G., Guattari, F.: What is Philosophy? Columbia University Press, New York (1994).
8. Dyson, F.: A meeting with Enrico Fermi. How one intuitive physicist rescued a team from fruitless research. Nature **427**(6972), 297 (2004). DOI 10.1038/427297a
9. Engelbrecht, J., Tamm, K., Peets, T.: On mathematical modelling of solitary pulses in cylindrical biomembranes. Biomech. Model. Mechanobiol. **14**, 159–167 (2015). DOI 10.1007/s10237-014-0596-2
10. Engelbrecht, J., Tamm, K., Peets, T.: Signals in nerves from the philosophical viewpoint. PhilSci:18135 [preprint] (2020)
11. Eringen, A.C., Maugin, G.A.: Electrodynamics of Continua I. Springer New York, New York, NY (1990). DOI 10.1007/978-1-4612-3226-1
12. Gavaghan, D., Garny, A., Maini, P.K., Kohl, P.: Mathematical models in physiology. Philos. Trans. R. Soc. A Math. Phys. Eng. Sci. **364**(1842), 1099–1106 (2006). DOI 10.1098/rsta.2006.1757
13. Heimburg, T., Jackson, A.D.: On soliton propagation in biomembranes and nerves. Proc. Natl. Acad. Sci. USA **102**(28), 9790–9795 (2005). DOI 10.1073/pnas.0503823102
14. McCulloch, A.D., Huber, G.: Integrative biological modelling *in silico*. In: G. Bock, J.A. Goode (eds.) *'In Silico'* Simul. Biol. Process., pp. 4–25. John Wiley & Sons, Chichester (2002). DOI 10.1002/0470857897.ch2
15. Noble, D.: Chair's Introduction. In: G. Bock, J.A. Goode (eds.) *'In Silico'* Simul. Biol. Process. Novartis Found. Symp. 247, Vol. 247, pp. 1–3. Novartis Foundation (2002). DOI 10.1002/0470857897.ch1
16. Tveito, A., Jæger, K.H., Lines, G.T., Paszkowski, Ł., Sundnes, J., Edwards, A.G., Mäki-Marttunen, T., Halnes, G., Einevoll, G.T.: An evaluation of the accuracy of classical models for computing the membrane potential and extracellular potential for neurons. Front. Comput. Neurosci. **11**, 27 (2017). DOI 10.3389/fncom.2017.00027
17. Weiss, J.N., Qu, Z., Garfinkel, A.: Understanding biological complexity: lessons from the past. FASEB J. **17**(1), 1–6 (2003). DOI 10.1096/fj.02-0408rev

Appendices

Appendix A
The Numerical Scheme

If you want to apply mathematics, you have to live the life of differential equations..

Gian-Carlo Rota, 2008

The pseudospectral method (PSM) (see [1, 8]) is used to solve the system of dimensionless model equations Eqs. (8.1) to (8.5) which is together with notations repeated here for the sake of readability:

$$
\begin{aligned}
&Z_T = DZ_{XX} + Z\left[Z - (A_1 + B_1) - Z^2 + (A_1 + B_1)Z\right] - J, \\
&J_T = \epsilon_1\left[(A_2 + B_2)Z - J\right], \\
&P_{TT} = c_2^2 P_{XX} - \mu_2 P_T + F_2(Z, J), \\
&U_{TT} = c_3^2 U_{XX} + NU U_{XX} + MU^2 U_{XX} + NU_X^2 + 2MUU_X^2 - \\
&\qquad H_1 U_{XXXX} + H_2 U_{XXTT} - \mu_3 U_T + F_3(Z, J, P), \\
&W = KU_X, \\
&\Theta_T = \alpha\Theta_{XX} + F_4(Z, J, P, U), \\
&\Omega_T = \zeta J - \epsilon_4\Omega.
\end{aligned} \tag{A.1}
$$

Here Z is the action potential, J is the recovery current, A_i, B_i are the 'electrical' and 'mechanical' activation coefficients, D, ϵ_1 are coefficients, $U = \Delta\rho$ is the longitudinal density change in the lipid layer, c_3 is the velocity of unperturbed state in lipid bilayer, M, N are the nonlinear coefficients, H_1, H_2 are the dispersion coefficients, μ_3 is a dissipation coefficient for the mechanical wave in biomembrane, P is the pressure, c_2 is the characteristic velocity in the fluid and μ_2 is the (viscous) dampening coefficient for the axoplasm, W is the transverse displacement and K is a coefficient. The notation Θ represents (local) temperature and α is a coefficient characterising the temperature diffusion within the environment. 'Mechanical' activation coefficients in the AP and ion current expressions are connected to the improved Heimburg-Jackson part (the LW) of the model as $B_1 = -\beta_1 U$ and $B_2 = -\beta_2 U$ where β_1, β_2 are the mechanical coupling coefficients. The coupling forces are the same as noted before in Eqs. (8.6) to (8.8). The last equation for the Ω is needed for the coupling force F_4 and it describes the evolution of an internal variable which was used for describing endothermic processes acting on a different (longer) timescale than the nerve signal ensemble. The ζ and ϵ_4 are coefficients.

© The Author(s), under exclusive license to Springer Nature Switzerland AG 2021
J. Engelbrecht et al., *Modelling of Complex Signals in Nerves*,
https://doi.org/10.1007/978-3-030-75039-8

A.1 Initial and Boundary Conditions

A $sech^2$-type localised initial condition with initial amplitude Z_o is applied to Z in system (A.1) and we make use of the periodic boundary conditions for all model equations

$$
\begin{aligned}
&Z(X,0) = Z_o \operatorname{sech}^2 B_o X, \ Z(X,T) = Z(X + 2Km\pi, T), \ m = 1, 2, \ldots, \\
&J(X,0) = 0, \ J_T(X,0) = 0, \ J(X,T) = J(X + 2Km\pi, T), \ m = 1, 2, \ldots, \\
&U(X,0) = 0, \ U_T(X,0) = 0, \ U(X,T) = U(X + 2Km\pi, T), \ m = 1, 2, \ldots, \\
&P(X,0) = 0, \ P_T(X,0) = 0, \ P(X,T) = P(X + 2Km\pi, T), \ m = 1, 2, \ldots, \\
&\Theta(X,0) = 0, \ \Theta(X,T) = \Theta(X + 2Km\pi, T), \ m = 1, 2, \ldots, \\
&\Omega(X,0) = 0, \ \Omega(X,T) = \Omega(X + 2Km\pi, T), \ m = 1, 2, \ldots,
\end{aligned}
\tag{A.2}
$$

where K is the number of 2π sections in the spatial period. The amplitude of the initial 'spark' is Z_o and the width parameter is taken as B_o. In a nutshell – such an initial condition is a narrow 'spark' in the middle of the considered space domain with the amplitude above the threshold resulting in the usual FHN action potential formation which then proceeds to propagate in the positive and negative directions of the 1D space domain under consideration. For all other equations, we take initial excitation to be zero and make use of the same periodic boundary conditions. The solution representing the mechanical and pressure wave is generated over time as a result of coupling with the action potential and ion current parts in the model system. Periodic boundary conditions are a requirement for using the PSM method.

It should be added that the action potentials annihilate each other during the interaction (see Fig. 6.3), as expected but the mechanical and pressure waves can keep on going through many interactions if one uses the fact that we have periodic boundary conditions for taking a look at the interactions of the modelled wave ensembles. However, as we have also the added 'friction' type terms in the relevant governing equations eventually the mechanical waves dissipate into the heat without the driving signal (the AP) providing them energy influx.

A.2 The Derivatives and Integration

For numerical solving of the system (A.1) the discrete Fourier transform (DFT) based (PSM) (see [1, 8]) is used. Variable Z can be represented in the Fourier space as

$$
\widehat{Z}(k,T) = F[Z] = \sum_{j=0}^{n-1} Z(j\Delta X, T) \exp\left(-\frac{2\pi i j k}{n}\right),
\tag{A.3}
$$

where n is the number of space-grid points ($n = 2^{12}$, for example), $\Delta X = 2\pi/n$ is the space step, $k = 0, \pm 1, \pm 2, \ldots, \pm(n/2 - 1), -n/2$; i is the imaginary unit, F denotes the DFT and F^{-1} denotes the inverse DFT. The idea of the PSM is to approximate space derivatives by making use of the DFT

$$\frac{\partial^m Z}{\partial X^m} = \mathrm{F}^{-1} \left[(ik)^m \mathrm{F}(Z) \right] \tag{A.4}$$

reducing, therefore, the partial differential equation (PDE) to an ordinary differential equation (ODE) and then to use standard ODE solvers for integration in time. The model (A.1) contains a mixed derivative term and coupling force terms can be taken either as a space derivative which can be found like in Eq. (A.4) or time derivative which is not always suitable for a direct PSM application and need to be handled separately.

For integration in time, the model system (A.1) is rewritten as a system of first-order ODE's after the modification to handle the mixed partial derivative term and a standard numerical integrator is applied. In the numerical examples given in the present book the ODEPACK FORTRAN code (see [3]) ODE solver is used by making use of the F2PY (see [7]) generated Python interface. Handling of the data and initialisation of the variables is done in Python by making use of the package SciPy (see [6]) and the numerical results are analysed and visualised in the Matlab environment.

A.3 Handling of Mixed Derivatives

Normally the PSM algorithm is intended for $u_t = \Phi(u, u_x, u_{2x}, \ldots, u_{mx})$ type equations. However, we have a mixed partial derivative term $H_2 U_{XXTT}$ in Eqs. (A.1) and as a result some modifications are needed (see [4, 5, 8]). Rewriting system (A.1) the equation for U so that all mixed partial derivatives including time are in the left-hand side of the equation

$$\begin{aligned} U_{TT} - H_2 U_{XXTT} = & c_3^2 U_{XX} + NUU_{XX} + MU^2 U_{XX} + NU_X^2 + 2MUU_X^2 - \\ & H_1 U_{XXXX} - \mu_3 U_T + F_3 (Z, J, P) \end{aligned} \tag{A.5}$$

allows one to introduce a new variable $\Phi = U - H_2 U_{XX}$. After that, making use of properties of the DFT, one can express the variable U and its spatial derivatives in terms of the new variable Φ:

$$U = \mathrm{F}^{-1} \left[\frac{\mathrm{F}(\Phi)}{1 + H_2 k^2} \right], \qquad \frac{\partial^m U}{\partial X^m} = \mathrm{F}^{-1} \left[\frac{(ik)^m \mathrm{F}(\Phi)}{1 + H_2 k^2} \right]. \tag{A.6}$$

Finally, in system (A.1) the equation for U can be rewritten in terms of the variable Φ as

$$\begin{aligned} \Phi_{TT} = & c_3^2 U_{XX} + NUU_{XX} + MU^2 U_{XX} + NU_X^2 + 2MUU_X^2 - \\ & H_1 U_{XXXX} - \mu_3 U_T + F_3 (Z, J, P), \end{aligned} \tag{A.7}$$

where all partial derivatives of U with respect to X are calculated in terms of Φ by using expression (A.6) and therefore one can apply the PSM for numerical integration of Eq. (A.7). Other equations in the model (A.1) are already written in the form which can be solved by the standard PSM. One should note that the dissipative term involving U_T was left on the rhs. Technically it is also possible to use U_X instead of U_T as in the 1D case the qualitative shape is the same, however, physically correct dissipative term involves the time derivative (U_T or U_{TXX}, etc). The calculation of U_T and P_T is outlined in the following A.4 section.

A.4 The Time Derivatives P_T, U_T and J_T

The time derivatives P_T, U_T (the lowest order dissipation in mechanical waves) and J_T (the coupling forces) are found using different methods. For finding P_T and U_T it is enough to write the corresponding equations in system (A.1) as two first-order ODE's which is done anyway for the numerical integration and it is possible to extract P_T and U_T from there directly

$$
\begin{aligned}
P_T &= \bar{P} \\
\bar{P}_T &= c_f^2 P_{XX} - \mu_2 P_T + F_2(Z, J), \\
U_T &= \bar{U} \\
\bar{U}_T &= c_3^2 U_{XX} + N U U_{XX} + M U^2 U_{XX} + N U_X^2 + 2 M U U_X^2 - \\
&\quad H_1 U_{XXXX} + H_2 U_{XXTT} - \mu_3 U_T + F_3\,(Z, J, P)\,.
\end{aligned}
\tag{A.8}
$$

For finding J_T a basic backward difference scheme is used

$$
J_T\,(n, T) = \frac{J(n, T) - J(n, (T - dT))}{T - (T - dT)} \approx \frac{\Delta J(n, T)}{dT},
\tag{A.9}
$$

where J is the ion current from Eqs. (A.1), n is the spatial node number, T is the dimensionless time and dT is the integrator internal time step value (which is varied) and in the examples given in the present book the integrator was allowed to take up to 10^6 additional internal time steps between ΔT values to provide the desired numerical accuracy. See the following technical details (Sect. A.5) for the notes on handling first time step and the last integrator internal time step (where the $dT = 0$).

A.5 Technical Details and Numerical Accuracy

As noted, the calculations are carried out with the Python package SciPy (see [6]), using the FFTW library (see [2]) for the DFT and the F2PY (see [7]) generated Python interface to the ODEPACK FORTRAN code (see [3]) for the ODE solver. The particular integrator used is the 'vode' with options set to nsteps= 10^6, rtol= $1e^{-11}$, atol= $1e^{-12}$ and ΔT varies but is in most simulations equal to one.

It should be noted that typically the hyperbolic functions like the hyperbolic secant $\text{sech}^2(X)$ in our initial conditions in (A.2) are defined around zero. However, in the numerical examples given in present book the spatial period is taken from 0 to $K \cdot 2\pi$ which means that the noted functions in (A.2) are shifted to the right (in direction of the positive axis of space) by $K \cdot \pi$ so the shape typically defined around zero is actually in our case located in the middle of the spatial period. This is a matter of preference (in the present case the reason is to have a more convenient mapping between the values of X and indices) and the numerical results would be the same if one would be using a spatial period from $-K \cdot \pi$ to $K \cdot \pi$.

The 'discrete frequency function' k in (A.4) is typically formulated on the interval from $-\pi$ to π, however, we use a different spatial period than 2π and also shift our space to be from 0 to $K \cdot 2\pi$ meaning that

$$k = \left[\frac{0}{K}, \frac{1}{K}, \frac{2}{K}, \ldots, \frac{n/2-1}{K}, \frac{n/2}{K}, -\frac{n/2}{K}, -\frac{n/2-1}{K}, \ldots, -\frac{n-1}{K}, -\frac{n}{K} \right], \quad (A.10)$$

where n is the number of the spatial grid points uniformly distributed across our spatial period (the size of the Fourier spectrum is $(n/2)$ which is, in essence, the number of spectral harmonics used for approximating the periodic functions and their derivatives) and K is the number of 2π sections in our space interval.

There are few different possibilities for handling the division by zero rising in Eq. (A.9) during the initial initialisation of the ODE solver and when the numerical iteration during the integration reaches the desired accuracy resulting in a zero-length time step. For the initialisation of the numerical function initial value of 1 is used for dT. This is just a technical nuance as during the initialisation the time derivative will be zero anyway as there is no change in the value of $J(n, 0)$. For handling the division by zero during the integration when ODE solver reaches the desired accuracy using values from two steps back from the present time for J and T is computationally the most efficient. Another straightforward alternative is using a logical cycle inside the ODE solver for checking if dT would be zero but this is computationally inefficient. In the present paper using a value two steps back in time for calculating J_T is used for all presented results involving J_T. The difference between the numerical solutions of the J_T with the scheme using a value 1 step back and additional logic cycle for checking in case of division by zero and using two steps back in time scheme only if division by zero occurs is approximately 10^{-6} and is not worth the nearly twofold increase in the numerical integration time.

The accuracy of the numerical solutions is approximately 10^{-7} for the fourth derivatives, approximately 10^{-9} for the second derivatives and approximately 10^{-11} for the time integrals (see, for example, [9] for details on the numerical accuracy of derivatives in PSM). The accuracy of J_T is approximately 10^{-6} which is adequate and very roughly in the same order of magnitude as the fourth spatial derivatives. The accuracy estimates are not based on the solving system (A.1) with the presented parameters and are based instead on using the same scheme with the same technical parameters for finding the derivatives of $\sin(x)$ and comparing these to an analytic solution. Also, it should be noted that in the PST the spectral filtering is a common approach for increasing the stability of the scheme – in the numerical simulations for the present book the filtering (suppression of the higher harmonics in the Fourier spectrum) is not used although the highest harmonic (which tends to collect the truncation errors from the finite numerical accuracy of floating-point numbers in the PST schemes) is monitored as a 'sanity check' of the scheme.

References

1. Fornberg, B.: A Practical Guide to Pseudospectral Methods. Cambridge University Press, Cambridge (1998)
2. Frigo, M., Johnson, S.: The design and implementation of FFTW 3. Proc. IEEE **93**(2), 216–231 (2005)
3. Hindmarsh, A.: ODEPACK, A Systematized Collection of ODE Solvers, vol. 1. North-Holland, Amsterdam (1983)
4. Ilison, L., Salupere, A.: Propagation of sech^2-type solitary waves in hierarchical KdV-type systems. Math. Comput. Simul. **79**(11), 3314–3327 (2009). DOI 10.1016/j.matcom.2009.05.003.
5. Ilison, L., Salupere, A., Peterson, P.: On the propagation of localized perturbations in media with microstructure. Proc. Estonian Acad. Sci. Physics, Math. **56**(2), 84–92 (2007).
6. Jones, E., Oliphant, T., Peterson, P.: SciPy: Open Source Scientific Tools for Python. http://www.scipy.org (2007).
7. Peterson, P.: F2PY: Fortran to Python interface generator. http://cens.ioc.ee/projects/f2py2e/ (2005)
8. Salupere, A.: The pseudospectral method and discrete spectral analysis. In: E. Quak, T. Soomere (eds.) Appl. Wave Math., pp. 301–334. Springer, Berlin (2009). DOI 10.1007/978-3-642-00585-5.
9. Tamm, K., Lints, M., Kartofelev, D., Simson, P., Ratas, M., Peterson, P.: Practical notes on selected numerical methods with examples. Tech. rep., Institute of Cybernetics, Tallinn University of Technology, Tallinn (2015). DOI 10.13140/RG.2.2.15205.81121

Appendix B
The example scripts

The purpose of computing is insight, not numbers.

Richard Hamming

This appendix contains two example scripts. First, one for numerical integration of the proposed model equations written in Python and second, a Matlab script for visualising the wave ensemble by taking the output file of the first script and plotting it. The resulting figure is included for the sake of completeness. One should note that in Python the index starts from zero while in Matlab the index starts from one. An open-source alternative to Matlab is GNU Octave.

B.1 Example: the script for an example

The following numerical script runs as presented (tested in 64bit branch of the Python 3.4.3 with SpiPy 0.16.1 and NumPy 1.9.3 under Windows 8.1 Pro) if the relevant Python and SciPy packages are correctly installed. The parameters included in the script lead to an emergence of a wave ensemble from an non-zero initial condition only for the FHN equation.

```
#coding=iso-8859-15
#/usr/bin/env python
#kert tamm , kert@ioc.ee
#-------------------------------------------------------------
import numpy
import scipy
from scipy.integrate import * #ode
from scipy.fftpack import * #diff
from scipy.io import savemat
from numpy.core import * #transpose
from time import *
import os

# functions
def cosh(x): # hyperbolic cosine
```

```
        internal = numpy.exp(x)/2+1/(2*numpy.exp(x))
        return internal

#scal wavenum, shift from -pi to +pi to 0 to l2pi * 2 pi
def oomega(n):
    qq = 0.0
    kk = 0
    while kk<(n/2):
        omega[kk] = qq/l2pi #first half
        omega[n/2+kk] = (-(n/2)+qq)/l2pi #second half
        qq=qq+1.0
        kk=kk+1
    return omega


# transform back to real space from Fourier space for U
# not used currently as rev transform moved to EQSnv
# left here as an example of standalone reverse transform
def ibioD(resultx,tvektorx,n):
    i = 0 ; um_t_real = [] ; aprox_vel = []
    temp2 = [] ; ru = [] ; fft_r = []
    temp3 = [] ; rut = [] ; fft_rt = []
    omega = oomega(n)
    while i<len(tvektorx): # transform to real space
        temp2 = resultx[i][2*n:3*n] # space
        temp3 = resultx[i][3*n:4*n] # velocity
        fft_r = fft(temp2,n,-1) # r forward
        fft_rt = fft(temp3,n,-1)
        r_temp = numpy.arange(n,dtype=numpy.complex128)
        rt_temp = numpy.arange(n,dtype=numpy.complex128)
        k = 0
        while k < n: # index 0 to n-1. reverse transform
            r_temp[k] = fft_r[k]/(1+HH2*omega[k]**2)
            rt_temp[k] = fft_rt[k]/(1+HH2*omega[k]**2)
            k = k+1
        ru = ifft(r_temp).real # end of reverse transform
        rut = ifft(rt_temp).real
        um_t_real.append(ru)
        # space [column = coordinate, row = time]
        aprox_vel.append(rut)
        i=i+1
    return um_t_real,aprox_vel

def Initconditions(Au,Av,Bo,x,l2pi): #Sech**2 Initial cond.
    # Au transmembrane potential amplitude,
    # Av recovery current amplitude,
```

```
      # Bo width parameter for intial sech2 pulse
      xx = x-12pi*numpy.pi #profile peak to the middle of space
      u0 = Au/(cosh(Bo*xx))**2 # AP initial profile
      v0 = Av/(cosh(Bo*xx))**2 # J initial profile or zeros(n)
      yf[:n] = u0              # Z (the AP)
      yf[n:2*n] = v0           # J (the ion current)
      yf[2*n:3*n] = zeros(n) # LW eq 1    U_T
      yf[3*n:4*n] = zeros(n) # LW eq 2    U(X,0)
      yf[4*n:5*n] = zeros(n) # PW eq 1    P_T
      yf[5*n:6*n] = zeros(n) # PW eq 2    P(X,0)
      yf[6*n:7*n] = zeros(n) # Temperature Theta(X,0)
      yf[7*n:8*n] = zeros(n) # First Internal variable
      yf[8*n:9*n] = zeros(n) # Second Internal variable
      return yf.real

  def EQSnv(t,yfhn): #FHNc+iHJc+WEqc+Temperature eq sys solving
      #backward difference scheme for some time derivatives
      global TTvec; global VTvec; global JTvect;
      global ZTvec; global ZTvect;
      nn = int(len(yfhn)/9); #DeltaVT=zeros[nn];
      ru = yfhn[:nn]; # ru - potential Z
      rv = yfhn[nn:2*nn]; # rv - ion current J
      TTvec[0] = TTvec[1]; TTvec[1] = TTvec[2];
      TTvec[2] = TTvec[3]; TTvec[3] = numpy.copy(t);
      DeltaT = TTvec[-1]-TTvec[-3];
      VTvec[0][:nn] = VTvec[1][:nn];
      VTvec[1][:nn] = VTvec[2][:nn];
      VTvec[2][:nn] = VTvec[3][:nn];
      VTvec[3][:nn] = numpy.copy(rv[:nn]);
      DeltaVT=VTvec[-1][:nn]-VTvec[-3][:nn]
      #calc -2 (1 step), -3 (2 steps) or -4 (3 steps)
      # ------------
      ZTvec[0][:nn] = ZTvec[1][:nn];
      ZTvec[1][:nn] = ZTvec[2][:nn];
      ZTvec[2][:nn] = ZTvec[3][:nn];
      ZTvec[3][:nn] = numpy.copy(ru[:nn]);
      DeltaZT=ZTvec[-1][:nn]-ZTvec[-3][:nn]
      #calc -2 (1 step), -3 (2 steps) or -4 (3 steps)
      # ------------
      v_t = numpy.copy(numpy.divide(DeltaVT[:nn],DeltaT)); #J_T
      z_t = numpy.copy(numpy.divide(DeltaZT[:nn],DeltaT)); #Z_T
      JTvect[:nn] = numpy.copy(v_t[:nn]); #storing  j_t
      ZTvect[:nn] = numpy.copy(z_t[:nn]); #storing  z_t
      hv = yfhn[3*n:4*n] # LW velocity
      P = yfhn[4*n:5*n] # pressure
```

```
    Pt = yfhn[5*n:6*n] # pressure velocity
    TMP = yfhn[6*n:7*n] #temperature evolution in time
    KCen = yfhn[7*n:8*n] #temp change from endothrm reaction
    KCex = yfhn[8*n:9*n] #temp change from exotherm reaction
# trasform for LW from Fourier space to real space ibioD
    temp2 = yfhn[2*n:3*n] #the LW for transf to realspace
    fft_r = fft(temp2,nn,-1) #r forward
    r_temp = numpy.arange(nn,dtype=numpy.complex128)
    kkk = 0
    while kkk < nn: #index 0 to n-1. for reverse transform
        r_temp[kkk] = fft_r[kkk]/(1+HH2*omega[kkk]**2)
        kkk = kkk+1
    hu = ifft(r_temp).real #end of reverse transform
    hu2 = hu * hu # U (the LW) squared
    u2 = ru*ru  # Z^2 (the AP)
    b1 = -beta1 * hu; b2 = -beta2 * hu; # pos LW damp AP, b_i
    u_x = diff(ru,1,period=2*numpy.pi*l2pi)  # Z 1 deriv
    u_xx = diff(ru,2,period=2*numpy.pi*l2pi)  # Z 2 deriv
    v_x = diff(rv,1,period=2*numpy.pi*l2pi)  # J 1 deriv
    v_xx = diff(rv,2,period=2*numpy.pi*l2pi)  # J 2 deriv
    hu_x = diff(hu,1,period=2*numpy.pi*l2pi)   # LW 1 deriv
    hu_x2 = hu_x * hu_x # LW square of the first derivative
    hu_xx = diff(hu,2,period=2*numpy.pi*l2pi)  # LW 2 deriv
    #hu_xxx = diff(hu,3,period=2*numpy.pi*l2pi)  # LW 3 deriv
    P_x = diff(P,1,period=2*numpy.pi*l2pi)  # P 1 deriv
    P_xx = diff(P,2,period=2*numpy.pi*l2pi)  # P 2 deriv
    #P_xxx = diff(P,3,period=2*numpy.pi*l2pi)  # P 3 deriv
    hu_xxxx = diff(hu,4,period=2*numpy.pi*l2pi)# LW 4 deriv
    TMP_xx = diff(TMP,2,period=2*numpy.pi*l2pi)  # TMP 2 deriv
    ytfhn[:nn] = ru*(ru-(a1+b1)-u2+(a1+b1)*ru)-rv+d1*u_xx
    ytfhn[nn:2*nn] = e1*(-rv+(a2+b2)*ru)+d2*v_xx
    ytfhn[2*nn:3*nn] = hv # d hu / dt = hv or u_t
    ytfhn[3*nn:4*nn] = c2*hu_xx+PP*hu*hu_xx+QQ*hu2*hu_xx +\
PP*hu_x2+2*QQ*hu*hu_x2-HH*hu_xxxx+gamma1*Pt+\
gamma2*v_t-gamma3*z_t-mu2*hv
    ytfhn[4*nn:5*nn] = Pt # d hu / dt = hv
    ytfhn[5*nn:6*nn] = cf2*P_xx+eta1*u_x+eta2*v_t+\
eta3*z_t-mu*Pt
    ytfhn[6*nn:7*nn] = Kt*TMP_xx+tau1*ru**2+tau2*(Pt+mu*P)+\
tau3*(hv+mu2*hu)-e2*KCen+e3*KCex
    ytfhn[7*nn:8*nn] = ct*rv-e2*KCen #tmp drop endo chem reac
    ytfhn[8*nn:9*nn] = ct*rv-e3*KCex #tmp gain exo chem reac
    return ytfhn


print('StartSolving')
```

```
# parameters and solving the system of PDE's
n = 2**11 #spatial grid points (for best performance use 2^n)
l2pi = 2*48 #number of 2 pi section in space
dt = 2. #time step of saving space vectors
tend = 770.1 # end time to integrate to
narv = int(tend/dt)#numbr of space vec to be saved in time
omega=numpy.arange(n,dtype=numpy.float64); omega=oomega(n);
x = numpy.arange(n,dtype=numpy.float64)*l2pi*2*numpy.pi/n #x
yfhn = numpy.arange(n*9,dtype=numpy.float64) #vect 9 eq
ytfhn = numpy.arange(n*9,dtype=numpy.float64) #vect 9 eq
yf = numpy.arange(n*9,dtype=numpy.float64) #vect 9 eq in row
# few variables for backward difference scheme for Zt and Jt
VTvec = numpy.zeros((4, n),dtype=numpy.float64);
TTvec = numpy.zeros((4,1),dtype=numpy.float64);
ZTvec = numpy.zeros((4, n),dtype=numpy.float64);
TTvec[1] = -1.; TTvec[2] = -1.;
# FHN (the AP) and iHJ (the LW) parameters
d1 = 1; d2 = 0.00 #diffusion, d2 is zero in normal FHN eq
# d1 is parameter D in Eqs A.1
e1 = 0.018; #timescale difference of Z and J
# e1 is param epsilon_1 in Eqs A.1
a1 = 0.20; a2 = 0.20 #nlin & coupling electr (b1,b2 mech)
# a1,a2,b1,b2 represent A_1, A_2, B_1, B_2 in Eqs A.1
PP = 0.05; QQ = 0.05; #iHJ (the LW) nonlinear parameters
# PP is N and QQ is M in Eqs A.1
HH = 0.2 # H1 dispersion parameter in iHJ model
# HH is H_1 in Eqs A.1
HH2 = 0.99 # H2 dispersion parameter in iHJ model
# HH2 is H_2 in Eqs A.1
c2 = 0.10; #dimensionless sound velocity in lipid bilayer
# c2 is parameter c_{3}^{2} in Eqs A.1
# parameters b1,b2 are defined inside EQSnv block
# coupling force and dissipation parameters
beta1 = 0.025; beta2 = 0.025; gamma1 = 1.e-2; gamma2 = 1.e-3;
gamma3 = 1.e-5; mu2 = 0.05;
# parameters beta1 and beta2 are beta_1,beta_2 after Eqs A.1
# parameter mu2 is mu_3 in Eqs A.1
# gamma1,gamma2,gamma3 are in force F_3 in A.1 (see Eq. 8.7)
# Pressure parameters
cf2 = 0.09; eta1 = 1.e-3; eta2=1.e-2; eta3 = 1.e-4; mu=0.05;
# cf2 is c_{2}^{2} and mu is mu_2 in Eqs A.1
# eta1,eta2,eta3 are in force F_2 in A.1 (see Eq. 8.6)
# Heat equation parameters
Kt = 0.05; tau1 = 0.0005; tau2 = 1.e-5; tau3 = 1.e-5;
e2 = 0.00125; e3 = 0.0001; ct = 5.e-3
```

```
# Kt is alpha in Eqs A.1
# tau1,tau2,tau3 are in F_4 in Eqs A.1 (see Eq. 8.8)
# e2, e3 are epsilon_{42}, epsilon_{41} (see Eq. 8.12)
# ct is zeta (internal variable coeff.) in F_4 (see Eq 8.11)
# Initial condition parameters
Bo = 1.; Au = 1.2; Av=0.005; # initial pulse param
#os.chdir('D:\\Book\\Appendix');
#os.getcwd(); #choosing a save dir in windows
#file name for storing the results for later analysis
fn ='TEST_1_%s_n%s_d1_%s_d2_%s_e1_%s_e2_%s_a1_%s_a2_%s\
_Au_%s_Av_%s_Bo_%s_N_%s_M_%s_H1_%s_H2_%s_cf2_%s_mu_%s_B1_%s\
_B2_%s_G1_%s_G2_%s_G3_%s_E1_%s_E2_%s_E3_%s_Kt_%s.mat'\
%(l2pi,n,d1,d2,e1,e2,a1,a2,Au,Av,Bo,PP,QQ,round(HH,2),\
   round(HH2,2),cf2,mu,round(beta1,3),round(beta2,3),\
   round(gamma1,3),round(gamma2,6),round(gamma3,6),\
   round(eta1,6),round(eta2,6),round(eta3,6),round(Kt,3))
# integration
tvektorx = []; resultx = [];
JTvec = numpy.zeros((1, n),dtype=numpy.float64);
JTvect = numpy.zeros((1, n),dtype=numpy.float64);
ZZTvec = numpy.zeros((1, n),dtype=numpy.float64);
ZTvect = numpy.zeros((1, n),dtype=numpy.float64);
yf = Initconditions(Au,Av,Bo,x,l2pi);
t0 = 0.0; t=0.0;
runnerx = ode(EQSnv)
runnerx.set_integrator('vode',nsteps=1e5,max_step=dt,\
                       rtol=1e-11,atol=1e-12);
runnerx.set_initial_value(yf,t0);
time_start = clock()
while runnerx.successful() and runnerx.t < tend:
    tvektorx.append(runnerx.t);
    resultx.append(runnerx.y);
    JTvec=numpy.append(numpy.copy(JTvec),numpy.copy(JTvect),\
        axis=0);
    ZZTvec=numpy.append(numpy.copy(ZZTvec),\
        numpy.copy(ZTvect),axis=0);
    print('FHNc+iHJc+WEq+TMP ',runnerx.t);
    runnerx.integrate(runnerx.t+dt);
time_end=clock();
time_cost=abs(time_start-time_end); #the calculation time
numpy.transpose(resultx);
i=0;
fhnu = numpy.zeros((len(tvektorx), n))
fhnv = numpy.zeros((len(tvektorx), n))
ihju = numpy.zeros((len(tvektorx), n))
```

```
ihjv = numpy.zeros((len(tvektorx), n))
nvP = numpy.zeros((len(tvektorx), n))
nvPt = numpy.zeros((len(tvektorx), n))
TMP = numpy.zeros((len(tvektorx), n))
KCen = numpy.zeros((len(tvektorx), n))
KCex = numpy.zeros((len(tvektorx), n))
while i<len(tvektorx): # separation of vectors
    fhnu[i] = resultx[i][:n] #transmembrane potential
    fhnv[i] = resultx[i][n:2*n] #recovery current
    ihju[i] = resultx[i][2*n:3*n] # deltaroo in iHJ model
    ihjv[i] = resultx[i][3*n:4*n] # deltaroo vel in iHJ
    nvP[i] = resultx[i][4*n:5*n] # pressure
    nvPt[i] = resultx[i][5*n:6*n] # pressure velocity
    TMP[i] = resultx[i][6*n:7*n] # temperature
    KCen[i] = resultx[i][7*n:8*n] # temp drop int var
    KCex[i] = resultx[i][8*n:9*n] # temp gain int var
    i = i+1
#(ihju,ihjv) = ibioD(resultx,tvektorx,n) #rev transform
numpy.transpose(fhnu); numpy.transpose(fhnv);
numpy.transpose(ihju); numpy.transpose(ihjv);
JTvector=numpy.reshape(JTvec,(size(tvektorx)+1,n));
ZTvector=numpy.reshape(ZZTvec,(size(tvektorx)+1,n));
Ttick = numpy.array(tvektorx);
#save in Matlab format for later analysis and visualisation
sona = dict([('fhnu',fhnu),('fhnv',fhnv),('J_T',JTvector),\
            ('x',x),('l2pi',l2pi),('Au',Au),('Bo',Bo),\
            ('Ttick',Ttick),('TInt',time_cost),('D1',d1),\
            ('D2',d2),('a1',a1),('a2',a2),('e1',e1),\
            ('beta1',beta1),('beta2',beta2),\
            ('gamma1',gamma1),('gamma2',gamma2),('Av',Av),\
            ('iHJu',ihju),('iHJv',ihjv),('P',PP),('Q',QQ),\
            ('H1',HH),('H2',HH2),('NVP',nvP),('NVPt',nvPt),\
            ('eta1',eta1),('eta2', eta2),('mu',mu),\
            ('cf2',cf2),('c2',c2),('Z_T',ZTvector),\
            ('eta3',eta3),('gamma3',gamma3),('tau1',tau1),\
            ('tau2',tau2),('TMP',TMP),('alpha',Kt),\
            ('mu2',mu2),('tau3',tau3),('e2',e2),('ct',ct),\
            ('KCen',KCen),('KCex',KCex),('e3',e3)]);
savemat(fn,sona,do_compression=1);
```

B.2 Example: the script for visualisation

The following Matlab script plots the normalised wave ensemble solution (see the
Figure given below) in a quarter-space for the final time step reached during the
numerical integration. It is included for the sake of completeness to make it easier
to check if the numerical solving script is working as intended.

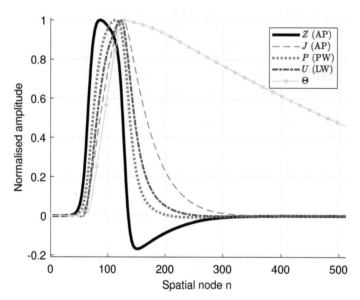

Fig. B.1 An example plot of a wave ensemble calculated with the numerical solving script presented
in the Section B.1.

```
% Book Appendix Figure Example
% kert@ioc.ee
% fhnu - FHN Z (the AP); fhnv - FHN J (the ion current)
% NVP - the pressure P (PW);
% iHJu - longitudinal density change U (the LW)
% TMP - the temperature change;
% Ttick - time vector; x - space vector
clear;
cd('D:\Book\Appendix') % set path to data file(s)
load(['TEST_1_96_n2048_d1_1_d2_0.0_e1_0.018_e2_0.00125'...
    '_a1_0.2_a2_0.2_Au_1.2_Av_0.005_Bo_1.0_N_0.05_M_'...
    '0.05_H1_0.2_H2_0.99_cf2_0.09_mu_0.05_B1_0.025_B2'...
    '_0.025_G1_0.01_G2_0.001_G3_1e-05_E1_0.001_E2_0.01'...
    '_E3_0.0001_Kt_0.05.mat'])
[m,n]=size(fhnu); % m time, n space
f1=figure(1); hold on;
```

```
% normalised result plots; index is (time,space)
plot(fhnu(end,:)'./max(abs(fhnu(end,:))'),'linestyle',...
    '-','color','k','linewidth',3);
plot(fhnv(end,:)'./max(abs(fhnv(end,:))'),'linestyle',...
    '--','color','r','linewidth',1);
plot(NVP(end,:)'./max(abs(NVP(end,:))'),'linestyle',...
    ':','color','m','linewidth',2.75);
plot(iHJu(end,:)'./max(abs(iHJu(end,:))'),'linestyle',...
    '-.','color','b','linewidth',2);
plot(TMP(end,:)'./max(abs(TMP(end,:))'),'linestyle',...
    '-','color','c','linewidth',1,'Marker','.',...
    'MarkerIndices',1:16:length(TMP(end,:)'),...
    'MarkerSize',12);
% plot quarter space showing the left propagating waves
axis([0 n/4 -0.21 1.01]);
grid on; set(gca,'FontSize',12);
ylabel('Normalised amplitude','FontSize',12);
xlabel('Spatial node n','FontSize',12);
l1=legend('$Z$ (AP)','$J$ (AP)','$P$ (PW)',...
    '$U$ (LW)','$\Theta$');
set(l1,'Interpreter','latex','FontSize',10,...
    'Location','NorthEast')
print(f1,'Fig_Example.eps','-depsc2','-painters');
```

Index

© The Author(s), under exclusive license to Springer Nature Switzerland AG 2021
J. Engelbrecht et al., *Modelling of Complex Signals in Nerves*,
https://doi.org/10.1007/978-3-030-75039-8